D0848180

Magnificent Failure

Magnificent Failure

free fall from the edge of space

Craig Ryan

Smithsonian Books
Washington and London

Copy editor: Therese Boyd
Production editor: Robert A. Poarch
Designer: Brian Barth

Ryan, Craig, 1953–
Magnificent failure : free fall from the edge of space /
Craig Ryan.
p. cm.
Includes bibliographical references and index.
ISBN 1-58834-141-0 (alk. paper)
1. Piantanida, Nicholas John, 1932–1966. 2. Balloonists—United States—Biography. 3. Parachuting—History. 4. Aeronautics—Records. I. Title.
TL540.P25 R93 2003
629.133'22'092—dc21 2002042615
[B]

British Library Cataloguing-in-Publication Data is available
Manufactured in the United States of America
09 08 07 06 05 04 03 5 4 3 2 1

∞ The paper used in this publication meets the minimum requirements of the American National Standard for Information Sciences—Permanence of Paper for Printed Library Materials ANSI Z39.48—1984.

For my father, Eugene Ryan

Adhere to your own act, and congratulate yourself if you have done something strange and extravagant, and broken the monotony of a decorous age.

—Ralph Waldo Emerson

The first object of the hero is to conquer himself.

—Juan Eduardo Cirlot

CONTENTS

ACKNOWLEDGMENTS

I was haunted by Nick Piantanida during the writing of my 1995 study of postwar high-altitude aeronautics, *The Pre-Astronauts: Manned Ballooning on the Threshold of Space*. I suspected there was more to his story than I'd been able to learn. Yet because of the way Piantanida's Project Strato-Jump had ended, because the aviation historians had so thoroughly ignored it, because much of the balloon community seemed to dismiss it, and because so little reliable information appeared to exist, I had given up pursuing the idea. So it's true to say that this book would never have happened had it not been for a phone call from Diane Shearin, Nick's middle daughter, in the fall of 1998. She wanted to learn more about her father. I hope that this book can be of some value to her in her quest to discover the father she never knew. Nick was an archetype, but he was also a complex individual—a complexity that is easy to overlook in the face of the stubborn singlemindedness that dominated the final three years of his life.

I was very fortunate to have the cooperation and able assistance of Vern Piantanida. Vern was always willing to talk about his brother and to answer hard questions. I appreciate the many hours he spent transcribing Nick's handwritten jungle logs and tracking down information on my behalf. In spite of the fundamental differences between these two brothers, knowing Vern has helped me to imagine what it might have been like to have known Nick.

Initially, Janice Post spoke with me a bit reluctantly and, I suspect, primarily out of courtesy. The story I wanted to tell was not one she was eager to relive. Eventually, in large part because of her daughters' encouragement, she agreed to share her memories with me and contributed aspects of the story that could have been provided by no one else. On occasion, it was painful. I am humbled by Janice's strength and by her unfailing grace.

Ed Yost did as much to help me understand Nick as anyone. While Ed's reputation as an inventor and a balloon operations expert was among the factors that convinced Nick to move his project to Sioux Falls in the winter of 1965/66, I'm convinced that it was Ed's spirit of adventure and his bedrock humanity that were decisive. Those qualities give Ed his unique insight into Nick's dream, and I thank him for sharing that insight with me. With so much misinformation in print about Nick Piantanida, I asked Ed why he chose to be so revealing to *me* so many years after the fact. "Well," he said in his best deadpan, one that suggests your question is either profound or just plain dumb, "you're the first writer who ever bothered to ask me what happened." As a rule, Ed doesn't think much of books like this one—mostly because he just knows too damn much. I will have to accept his judgment.

Yost's colleague at Raven Industries, Jim Winker, was a monumental help. Jim is an engineer by temperament as well as by profession. He doesn't miss much and he doesn't mess around; he keeps good records and he remembers things. And remembering events that happened nearly forty years ago is, as I was regularly reminded during my research, an uncommon gift. Jim read early drafts of the manuscript and offered many suggestions, often supported by research materials I might never have found without him. I appreciate his patience, his skepticism, and his insistence on accuracy, even when it threatened to undercut drama.

Roger Vaughan was perhaps the most objective of the "inside" observers of Project Strato-Jump. While he had grown close to Nick, and while his magazine had obtained exclusive rights to much of the story, he was not employed by the project or any of its contractors. His willingness to discuss those days was important to my understanding of the events he covered, and some of his interpretations and conclusions were clearly influential on my own.

I want to thank the marvelous Union City gang who grew up with Nick and who did their best to recreate those days for me: Fred Cranwell, Jim Lagomarsino, Ed Madsen, Bob Santomenna, and Bob Taglieri. Spending time with these guys is about as good as it gets.

The following people, as well as a number already mentioned, invited me into their lives and homes, and I thank them all for their hospitality and their help: Frank Heidelbauer, Jacques and Felicia Istel, Christine Kalakuka and Brent Stockwell, Debbie Piantanida, Paula Piantanida, Don Piccard, Russ Pohl, David Post, Karl and Lucy Stefan, and Marlene Winker.

Phil Chiocchio was an especially valuable source of information on Nick's early parachuting days as well as on all three Strato-Jump flights. Not only did he provide me with a copiously annotated copy of Barry Mahon's rare film *The Angry Sky*, but he contacted a number of members of the parachuting community and conducted initial interviews for me, notable among them Jim Bates, Bob Blanchard, Debbie Foster, Norm Heaton, Bill Jolly, Chuck MacCrowne, Dan Poynter, and Mike Turaff. Finding Phil was one of those lucky breaks a researcher can't create but can only stumble onto.

I want to offer special thanks to the great pilot, balloonist, and American hero whose record Nick wanted so badly to break: Joe Kittinger. Joe never met Nick face to face and is critical of the planning and the approach of Strato-Jump, but he went out of his way to help me whenever I asked. Joe knows more about manned high-altitude ballooning than anyone alive, and his cooperation made this a better book.

The following people aided me in my research, and I thank them all: Tracy Barnes, Earl Clifford, Tom Crouch, Lee Guilfoyle, Dick Kelly, John Kittelson, Kent Morstad, Becky Pope, Ben Raiche, and Al Tomnitz. Chris Johnsen supplied background information on decompression sickness, Sharon and Bob Glaeser shared their parachuting expertise, and Scott Ellsworth provided assistance on a number of historical matters. I am indebted to Jack Bassick at the David Clark Company for trusting me with the company files and for helping me contact others in Worcester who had been around in 1965-66, notably Tom Smith and Joe Ruseckas. Joel Strasser deserves special recognition for graciously allowing his superb photographs of the preparations for the second Strato-Jump launch to be used in this book, as does Jon Lee for his excellent artwork and for his insightful comments on the manuscript. I am also indebted to Olga Levadnaya who provided the translation of the article from the Russian periodical *The Aeronaut* (see bibliography). A special thanks to friends who took an interest: Ron and Julie Byers, Cory Carpenter, Richard Causby, Betsy Ellsworth, Patrick Hampton, Constance Harris, Robert Sleator, and Jim Stewart.

I'd like to acknowledge two individuals who died while this book was in progress. They both meant a lot to me. Suzanne Robinson Yost (1932–2001) was an extraordinary woman and a long-time stalwart of the ballooning community who brought a cheer and dignity to every room she ever entered. She is missed by many. Dr. John Paul Stapp (1910–99), the visionary behind the Air Force's great manned balloon programs and the founder of modern vehicle-safety research, devoted his life and uncompromising genius to the cause of humanity.

Thanks to Mark Gatlin and everyone at Smithsonian Books, and—once again—to my editor, Therese Boyd. Also to Andrea Carlisle for her expert eyes, literary acumen, and helpful suggestions. And a special thanks to my agent, Al Zuckerman.

Finally, I want to acknowledge my wife, Kathy Narramore, who has lived with this project for three years and whose myriad contributions to the book have been as valuable as her support for the author.

PROLOGUE

Somewhere You Couldn't Follow

He was a hometown legend.

"Not a good player," a former running mate insists, "a *great* player. Best I ever saw." Folks in the neighborhood still talk about getting Union City to rename Manhattan Avenue—a shaded, two-block street on the bluff high above Hoboken—Piantanida Avenue. Nick used to shoot baskets there. Somebody had nailed a rim to a telephone pole. The pole survives, listing slightly toward the Hudson River. The rim is long gone.

"You never saw him without that smile," an old friend tells me. "Everybody loved Nick."

But another guy, lowering his voice, makes a point of taking me into his confidence: "You know, I don't think any of us ever really knew Nick. Not really."

You listen long enough and it dawns on you: Here was a man without heroes. He had only himself. You'd be talking to him one minute and turn to find him gone. Or his eyes would drift and you'd know he was somewhere else. Somewhere you couldn't follow. Over the course of his life he had only a few close friends.

"I've heard him called a rebel," one of the guys from the neighborhood counsels me. "But Nick was never rebelling *against* anything. Nick was too busy. He had his own stuff going. And that's where he was most of the time."

Somewhere you couldn't follow.

"Nick was . . . Nick," they keep telling me with their Jersey shrugs, tired of trying to explain. I don't know what that means, exactly. But I'm learning.

Where is the line between courage and folly? What is the cost of a dream?

Beginning shortly after that autumn day in 1783 when the world's first pilot, François Pilâtre de Rozier, rose 3,000 feet beneath a cloth bag full of smoke and hot air and looked down at Paris from on high, history records an incessant assault on the altitude record, leading humankind finally beyond the bonds of planet Earth itself. With better equipment, better technique, better planning, a better grasp of the peculiar problems involved in keeping a human being safe and productive in the weird and hostile territory of the extreme skies, it would always be possible to fly a few feet higher than the fellow who came before. And given favorable conditions, a whole lot of nerve, plenty of money, and a little bit of luck, why couldn't you push a few feet higher yet?

By the spring of 1961, with the demise of a trio of ambitious scientific balloon programs run by the U.S. armed forces which had produced an ascending string of altitude records, the highest manned flight belonged to Comdr. Malcolm Ross and Lt. Comdr. Victor Prather of the U.S. Navy: 113,740 feet—nearly twenty miles—above the sea level of Earth. But because that great flight had ended tragically with Prather's death, and because manned ballooning had by that time slipped into near-total obscurity behind the brilliant glare of NASA's Project Mercury, the future of high-altitude aeronautics was very much in doubt. Many wondered whether funding could ever again be found for manned lighter-than-air exploration of the upper stratosphere. Persuasive voices like those of Ross, the Air Force's visionary Col. John Paul Stapp, and the German immigrant balloon-builder Otto Winzen argued for a role for the piloted gas balloon as a space-equivalent research platform that could save the research-and-development efforts of NASA—not to mention the nation's taxpayers—many millions of dollars in the design of vehicles and survival systems for the journey to the moon. But no one was listening.

Then, out of the wild blue, the manned balloon altitude mark was shattered. On the second day of February 1966 Nick Piantanida flew a full two miles beyond Ross and Prather. And he did it mostly on his own. He raised his own funds and planned his own project. He recruited his own staff and—

both for better and for worse—called his own shots. The new altitude mark was never officially recognized since the Fédération Aéronautique Internationale, the organization that sanctions and validates such feats, requires that the pilot land with his balloon, something Nick had never intended to do. Yet the accuracy of the altitude his team reported has never been questioned.

Nick flew higher than any balloonist before him. His unofficial record has endured for thirty-seven years and counting. It is the longest-standing absolute altitude record in the history of manned lighter-than-air flight.

Yet by the spring of that same year, both fate and the times had caught up to Nick Piantanida. The changes—in space technology and in America itself—would in short order erase Project Strato-Jump from memory. Because the thrust and focus of this tremendously difficult, perhaps quixotic, effort had never been about ballooning, but rather about parachuting, and because the project had come up short of its ultimate goal—to set a new world free-fall record—it is judged a failure.

Why read—indeed, why write—the history of a failure? Musty magazines and blurry microfilm record the tales of those who *almost* achieved the extraordinary. Books are reserved for triumphs and for the triumphant. History is the victor's story.

So why an examination of this particular failure? First, Project Strato-Jump was far from a total loss. It launched the world's highest manned balloon flight and returned the pilot safely to Earth. This achievement in itself demands a place of honor in the historical record, a place that it has so far been denied. Second, Strato-Jump serves as an enduring cautionary tale for adventurers of all stripes who aspire to accomplish those feats that have eluded their predecessors. There is much to be learned. And finally, the story of Project Strato-Jump is very much a tale of its time, and one peopled with a cast of remarkable individuals, American originals all. It is an inspirational story in some respects, a sad, troubling one in others. And it is a story that has yet to be widely—or very accurately—told.

Project Strato-Jump represents the last gasp, the very dead-end of stratospheric manned ballooning in the twentieth century. No human being has since taken a balloon into the space-equivalent conditions beyond our atmosphere. The year 1966, then, marks the end not just of a memorable man, but of a magnificent American dream, of an era.

Nick Piantanida's life was a legend, but no less his death. The matter of what transpired more than thirty-five years ago in the stratospheric near-vacuum above southern Minnesota has acquired the dizzying qualities of a *Rashomon*-like puzzle drama in which objective reality recedes in the face of endless individual perspectives. A dozen intersecting—often conflicting—theories have been advanced. A dozen more rumors have floated from drop zone to drop zone. Who was Piantanida? What drove him? And finally, what really happened to him? Even those who knew him best—his family, his friends, his colleagues—are mostly at a loss. "Nick was . . . Nick."

But history abhors a mystery. If we can't follow Nick, we can at least trace his tracks. We can revisit his world. And if we can assemble and examine the evidence (some of which has only recently been uncovered), if we can find a way to evaluate the accounts of the men and women who were there—those who knew Nick, those who worked with him, those who opposed him as well as those who shared his dream—it will be possible, finally, to say: This is it—the *real* story of Nick Piantanida and Project Strato-Jump.

There can never be another quite like it.

1 "SOMETHING IN THE AIR"

On a September morning in 1963, just after dawn, Nick Piantanida appeared in the office of the Lakewood Sport Parachute Center on the Jersey shore and counted out the thirty-five-dollar fee. A jungle bushwhacker by passion and exotic pet dealer by trade, he sat impatiently through the hour-long ground-school session that was designed to calm fears and bolster confidence. Physical fear was not much of a concern—Nick, at thirty-one years of age, felt himself, rather matter-of-factly, to be invincible—and confidence was already available to him in such copious quantities that his only real curiosity here was in more technical matters: the harness rigging, body position, velocity calculations, and landing posture. The hands-on portion of the training program was of more interest. Upon being issued a white jumpsuit, a hard-shell helmet, and a parachute rig known as a "gutter" because of the pillow-sized reserve-chute pack that was positioned in front, on top of the gut, Nick learned to step off of a three-foot platform and execute a textbook PLF (parachute landing fall): arms up to simulate gripping the risers, legs together, eyes on the horizon, rolling in the direction of the fall to distribute the force of the impact over the entire body. Having been a licensed multiengine pilot for more than a decade, he was comfortable with small airplanes—he'd owned a couple—and with heights. He remarked casually to the Lakewood staff that he'd been "getting around to" this parachuting thing for a while now.

After cinching down his rig, Nick strolled out to the old Noorduyn Norseman jump plane waiting on the ramp. He moved in a slow, rolling, slightly pigeon-toed gait that reminded some of the Lakewood regulars of John Wayne.

Following the grinding climb up to altitude, the pilot throttled back the engine and the instructor popped open the plane's door to establish visual contact with the drop zone and to properly account for the wind prior to spotting the jump. Nick smiled as he felt the rush start in his bloodstream. Moments later, there he was: stepping off the Norseman's strut on a static line at 3,000 feet, with just a breathtaking second of free fall before the canopy snapped open with a wallop and silenced the roar of the wind in his ears.

It wasn't at all what he'd imagined, although he admitted later that, in spite of his bluster, the moment of truth had briefly scared the hell out of him. He'd been anticipating a sensation of tremendous speed and the falling-elevator tug in his stomach, but he got neither. Instead he found himself sailing like some sky god through a strange and unexpectedly buoyant world with the entire wide universe tight in his hands, drifting in slow motion, poised and floating on the air. This was something beyond experience, wholly new and glorious: beyond mere fun, beyond even thrill. As if he'd somehow fallen smack into his own life.

After gliding down to the drop zone as he'd been taught and executing the landing, which was about like they'd told him it would be, like running off the edge of a tall table, Nick got to his feet and gazed back up at the late summer sky. The center's "desert rat," a high-school boy whose job it was to snag a rookie jumper's chute, hustled up and helped the new guy ease out of the nylon harness. As he watched the kid field-pack the parachute for return to the rigging loft, Nick, still buzzing, considered for a moment what might be possible. He'd once pioneered the first ascent of a particularly difficult cliff face deep in the jungles of South America. He'd helped build the great Verrazano Narrows Bridge. He'd handled king cobras with his bare hands. In spite of his relaxed bearing, there was a carnivorous energy to Nick, an intensity that seemed to percolate within him, just below the surface. He was an imposing figure, more than six feet tall and over 200 pounds, quick and powerful, with a meaty jaw, intense, hazel eyes, close-cropped, dark-brown hair, broad shoulders, and rippled biceps.

Nick had always been good at things, even better when people made the mistake of trying to convince him those things were impossible. Now he found himself in a new and unfamiliar world, wondering just what *was* possible here. How high could you go? And the complementary question: how far could you fall?

If it would ever be possible to trace what happened later back to a precise moment, it would be to the afterglow of that first jump at Lakewood. That may be as close as we can get to pinpointing the instant when converging elements began to form the vortex of imagination and will that would ultimately inhale money, mass media, and some of the most talented and knowledgeable individuals in mid-twentieth-century American aeronautics. The power of that vortex would continue to astonish and challenge others—maddeningly, in some cases—for decades. But it's not likely that it ever surprised Nick Piantanida himself.

Recreational parachuting in the early 1960s was a poor cousin of today's chic-risk pastime where soccer moms line up for weekend tandem jumps and pay fifty dollars extra for videotape souvenirs. Using patched-up Army surplus chutes, the old round ones that offered only crude steering and dropped a jumper like a side of beef (the rare stand-up landing was practically a cause for celebration), the early sport jumpers would tolerate all manner of terror and pain for the sake of the incomparably pure adrenaline rush of free fall. Once they'd tasted thirty seconds of free flight, they were drawn inexorably back to the drop zone for another fix. They learned and perfected the tricks of their trade mostly by painstaking trial and error. They sprained ankles and necks, broke ribs and wrists on hard landings. They whacked skulls and jaws as they attempted to link up in midair formations, sometimes spearing each other broadside when their goggles fogged up. All the jumpers' crotches and armpits were bruised where their harnesses yanked and sawed at the chute's opening shock. On a windy day, a jumper's greatest challenge could be to quickly dump the air out of his chute upon landing to avoid a nasty parasail ride into the roots and stumps at the edge of Lakewood's pine forests. On occasion they "impacted" the ground—as the accident reports put it—at somewhere in the neighborhood of 120 miles per hour. In jump parlance, this terminal maneuver was known as a "bounce."

The first World Parachuting Championship had been held in Bled, Yugoslavia, in 1951 and had gone almost unnoticed by newspapers across the Atlantic. Not a single American competed. While thousands of paratroopers had returned to the States following World War II with significant experience—an adventurous few even jumping for kicks as civilians after arriving stateside—bailing out of perfectly good airplanes was still deemed far too dangerous an activity for general recreation, much less upon which to organize a legitimate sport. The Civil Aeronautics Board, which considered parachuting strictly a lifesaving emergency procedure for aviators in distress, even found it necessary to issue a regulation prohibiting delayed-opening jumps at domestic air shows.

Then, in 1955, naturalized citizen Jacques-André Istel, a Princeton graduate and an investment banker by trade if not by inclination, returned to his native France to study the techniques of stable, controlled free fall. In the early days, the jumper leapt or fell away from the aircraft and literally cast his fate to the wind, tumbling randomly through the air and trusting his parachute to open without snagging a flailing limb. And the chutes did open safely . . . most of the time. The French had a term for these dervishlike descents: *making a mayonnaise*. The discovery of deliberate, face-to-earth free fall, in which the jumper assumes a spread-eagled and arched aerodynamic body position that provides not only a clean glide path but the ability to execute turns and more complex maneuvers, brought a predictability and increased safety to parachuting. It also provided the opportunity for skill, creativity, and—ultimately—competition.

Istel had always been something of a visionary. He came back from Europe with big ideas and, initially with the able assistance of American parachuting prodigy Joe Crane, began training what would become the nation's first precision jump team to compete on the international stage. These efforts led in short order to the formation of European-style parachute clubs at Ivy League colleges and elsewhere in the northeastern United States. The following year, Istel led the fledgling U.S. team to an impressive sixth-place finish at the world championships in Moscow.

At the time of the Moscow event, there were fewer than a dozen proficient free-fall parachutists in the entire United States, but that number increased rapidly in the ensuing years. In 1962, after badgering both the White House and Congress for funding, and after threatening to boil a live dog for TV news cameras in Boston Commons unless the Massachusetts state legislature made good on its financial commitment, Istel spearheaded the movement that brought the championship meet to the United States for the first time. Twenty-six countries competed in Orange, Massachusetts, with Americans winning both the men's and women's individual titles, as well as the women's team title. The men's team came in second just behind the perennial favorite Czech team. Notice was formally served that the sport of free fall had touched down on American soil. Istel went on to modify the charter of what had been the National Parachute Jumpers-Riggers to establish the first national organization dedicated to the pursuit of free fall. He called it the Parachute Club of America (which later became the U.S. Parachute Association). Istel and some of his parachuting cronies on the East Coast then went into business. They called their enterprise Parachutes, Incorporated, and began sponsoring commercial facilities like the popular jump center in Orange and its counterpart on the Jersey shore an hour south of New York City.

The Lakewood Center, just a few miles from the ocean, a nondescript compound of low, white clapboard buildings surrounded by sandy soil, pine trees, and rail fences, opened next to a small airfield early in the summer of 1963. The drop zone itself was a sixty-acre patch of ground that had been rototilled into the consistency of dirty beach sand. The Lakewood regulars, like most of the early jumpers elsewhere, were of a more-or-less common profile. In spite of a family atmosphere that regularly included wives, young children, and dogs, the jumpers at Lakewood were a cocky, tough, hard-drinking lot given to practical jokes. And regardless of where they'd come from, they had all taken the paths-less-traveled to get to Lakewood. It was a tight-knit group that crowded into small airplanes together, forced by circumstance to develop the trust necessary to maneuver in close proximity at terminal velocity thousands of feet above the ground. In these days before the advent of square "ram-air" chutes, before rip-stop fabrics and Velcro were commonplace, equipment malfunctions were relatively frequent. But the jumpers looked out for each other, bound by a mutual respect and a passion for the freedom of the skies. And no airmen were so free as those who flew like the birds, governed only by gravity and protected only by flimsy disks of silk or nylon. They hung together because the uniqueness of their enterprise—the heightened consciousness experienced in free fall, the very real danger of sudden death, the ego blast that came of embracing oblivion—isolated them from the everyday world of work and leisure. Jumping for them was nothing so mundane.

Free fall became quite literally a way of life.

Nick Piantanida returned to Lakewood every chance he got, even in the wind and rain when it was impossible to jump and there was nothing to do but sit around the drop zone with other hardcores and toss pea gravel from the landing pit at Coke bottles, telling stories and drinking beer. Beer and stories, he came to learn, were essential components of the jumper's life. Before the leaves were off the trees that fall, he was off the static line and free-falling. It was an exhilarating time. From 7,000 feet the free-fall duration was a full thirty seconds, enough time to link up with other jumpers and to pass objects back and forth. Nick smuggled a ten-foot python aboard one afternoon and he and his fellow jumpers passed the snake from hand to hand in midair.

They were calling the new sport "skydiving," and it was beginning to catch the attention of a curious public. Reporters and photographers were regular visitors to Lakewood where the last of the giant blimps from the Lakehurst Naval Air Station to the south were still common sights above the airfield. But what the Lakewood regulars could never adequately convey to the press was the sheer visceral experience of extravehicular flight. The sweet shock of the rush. You had to jump if you wanted to understand, they said. To the jumpers it was strictly an us-and-them world. If you weren't willing to jump, you were them. You were irrelevant.

By winter, an idea—a waking dream, really—had begun to form in Nick's imagination. It was a dream of the greatest parachute jump of all time. He was not a humble man, and this was the way his mind worked. He would find out who had made the highest jump in history, how high he'd jumped, and he would surpass it. It would take a monumental effort, he was sure, but he had little doubt that he was up to the task both physically and mentally. And he had a clear vision of himself standing as a colossus at the feat's conclusion.

At first he kept the dream to himself. As plans formed and mutated and re-formed in his mind, he confided finally in his wife. It was appropriate that she heard it first because hers would become a burden every bit as great as

Nick's. He was understandably reluctant to mention it to his new friends at Lakewood. He worried in the beginning about offending his more experienced brethren with the sheer audacity of the idea. These were men he respected and whose help he was surely going to need before all was said and done. But he had, through subtle inquiries, managed to identify his competition. He learned that only the previous November, a Russian soldier had survived a world-record jump from an altitude of 83,524 feet—almost fifteen miles above Earth.

Nick was never easily impressed, but this gave him pause. Fifteen miles! Where in the world did you find an aircraft to get you up to such a height? And what was the environment like at 80,000 feet? Would you need some kind of spacesuit? And how long would it take you to fall back to Earth? He could imagine a few of the special problems such a jump might present. Of course, with the Cold War in full freeze, the fact that the record was owned by a Russian made the dream just that much more delicious to contemplate. The Soviet Union had been kicking America's collective butt in space since the flight of *Sputnik* five years earlier. This would be a golden opportunity for Yankee revenge. With resolve and guts as his principal assets, Nick would personally settle all the scores.

The dream took on a palpable momentum as he began to experiment with free-fall techniques, teaching himself to spin, somersault, and knife through the air at speeds approaching 200 miles per hour. He was going to do this thing, he decided, or he was going to die trying. He felt confident that he had finally, after years of searching, stumbled onto a challenge great enough to sustain him. He had found the ultimate place where no one could follow. In some ways it came as a profound relief. He'd experienced misery and he'd known heartbreak, but his greatest fear was of inconsequentiality, the life of quiet desperation. Now here was his ticket, his deliverance from anonymity.

Nothing less than the course of the rest of his life had been set.

New World, New Jersey

Nicholas John Piantanida had been born on the Jersey side of the Hudson River on August 8, 1932. Within weeks, the family moved to the Bergen Beach section of Brooklyn, but a year later returned to New Jersey, settling into a house on 17th Street amid the immigrant bustle and grit of Depression-era Union City. In 1935 they moved in with Nick's maternal grandparents in an Italian neighborhood on Third Street when the old folks nearly lost their house for delinquent property taxes.

The Piantanidas had come to America from an island off the Dalmatian Coast of Croatia (at that time, Yugoslavia) where both of Nick's parents had been born. Korcula, the supposed birthplace of the explorer Marco Polo, was a picture-postcard jewel in the Adriatic Sea that had been occupied by nearly every European power at one time or another. The inhabitants earned their modest livings from fishing and agriculture, grapes and olives mostly, and from the quarrying of the region's resplendent white marble. The men hunted jackals in the hills for sport. But the unique specialty and identity of Korcula lay in their wooden boats—Korculans were master shipbuilders who rivaled even the great naval craftsmen of Venice. As the need for wooden vessels declined in the early twentieth century, the island found itself competing for the tourist trade. World War I brought economic catastrophe that forced mass emigration, much of it across the Atlantic to the United States.

Florenz Piantanida had been born within Korcula's ancient city walls in 1906, the youngest of thirteen children. One brother had already left for America and had settled in Hoboken, a second brother alighted in Jersey City, and a third had chosen Brooklyn. At age seventeen, Florenz disembarked at Ellis Island with his parents and became, for a while, a small boat builder in the New World. When the Great Depression hit, he took a job with a bank restoring repossessed houses. Then, just before the outbreak of World War II, he hired on at Todd Shipyard where he would work, first in Hoboken and later in Brooklyn, until his retirement in 1971.

Catherine Zarnecich's parents had lived within easy walking distance of the Piantanida home in Korcula. Her father left first for the United States and sent for his wife in 1911. When she became pregnant with Catherine she misread the symptoms and thought she was dying, occasioning a return to Korcula. After the birth on New Year's Day, 1913, mother and daughter returned to Hoboken where the family moved into a modest house just a few doors down from a Sicilian-born fireman named Sinatra whose son Francis would later make something of a name for himself as a singer.

The industrial hive of northern New Jersey was a far cry from the picturesque beauty of the Adriatic, but it held its own distinctive brand of urban charm, especially for young people. Joan Doherty Lovero touches on this

The Piantanida boys: Vern (*left*) and Nick. (Courtesy Vern Piantanida)

in her history *Hudson County: The Left Bank:* "Perhaps the children enjoyed life in Hudson County most of all. They could sneak onto ferries and ride to New York. They could prowl around the outskirts of the rail yards behind the backs of the railroad bulls and observe tugboats easing their vessels into berths. They could watch horses being shod and see movies being made. Innocent of environmental worries they could even swim in the Morris Canal. It must have been terrific fun."

Florenz first met Catherine at a dance at the Dalmatia Club in Hoboken, a social club for Croatian immigrants, in early 1931. They were married several months later at the Zarnecich home up on the bluff in Union City, a municipality created just six years earlier when Union Hill had been merged with West Hoboken.

Catherine would pass on her keen intelligence to her offspring, while Florenz bequeathed an estimable physical strength and stamina. He intentionally neglected, however, to pass on his trade skills to his firstborn, hoping—in a classic evocation of the immigrant's dream—that Nick would forego manual labor for a college education.

A second son, Vernal, followed five years later in January of 1938. The Piantanida boys were active, handsome kids with light-olive skin and dark hair who would develop almost mirror-opposite personalities, reflecting the schizophrenia of the nation's postwar psyche. Nick became the impulsive extrovert snatching for the freedom at the heart of the American dream.

Vern, on the other more practical hand, would buckle down and apply himself with monklike discipline to each endeavor he undertook. They were simply alternate manifestations of the same dream. Yet in spite of their close relationship and a deep mutual respect, the brothers were destined never to fully understand each other.

You truly were lucky to be a kid in Union City in the 1940s and early '50s. The area had been hammered hard by the Depression, but World War II had reinvigorated the entire county's economy, starting with the ports and shipyards. The young were nearly all first-generation Americans and nobody doubted that anything short of the sky was the limit. Proud, hardworking parents who in some cases could barely read English preached the values of education, sacrifice, compassion, devotion, and citizenship and sent their children out into an urban frontier full of hope and confidence.

Things were rarely dull. A veritable storehouse of raw ingredients for America's melting pot, on its high ground within sight of both Ellis Island and the Statue of Liberty, it was where the emissaries of the nations of Europe and beyond alighted and first engaged the social and cultural mathematics of human reinvention. First the Germans and the Irish, then the Italians and the Jews, then waves of Greeks, Poles, Armenians, Syrians, Slavs—and later Puerto Ricans and Cubans, Dominicans and Colombians. Union City was an incubator where the first generation rose up, got its bearings, and then moved on to make way for new immigrants who would repeat the cycle. And while they could see poverty if they looked around them—overcrowded cold-water flats and families of gypsies sleeping in storefront windows—most of the denizens of Third Street and the immediate environs felt little of hardship's sting themselves.

One of the kids who grew up with Nick in the neighborhood later described the essence of the place as some mysterious elixir. "There was something in the water," he said, "something in the air. It was a *great* place to grow up."

Nick was still a skinny kid of ten when he clambered up onto the roof of the yellow, two-story rowhouse on Third Street to conduct the first in a series of experiments in aerodynamics and mechanical force. In one hand he clutched a small parachute constructed from an old pillowcase and some twine, in the other a stray cat he had caught down by what the neighborhood kids called Scumbag Creek. The first trial drop was a bust. The parachute failed to open and the shanghaied test pilot made the ultimate sacrifice on the concrete below. Nick reshaped the canopy, caught another cat, and attempted to induce some lift by swinging the assembly, chute and startled cat, high in the air—but with the same fatal result. A brief flare of the canopy had done nothing to substantially diminish the impact. Clearly, there were engineering problems to be resolved. Nick worked with the chute, as the neighborhood cats presumably laid low. But nothing in Union City stayed quiet for long, and Nick suffered a fate only slightly less severe than that of the cats when Catherine found out about the feline slaughter on her property.

Not that any amount of punishment, corporal or otherwise, could discourage Nick for long. For the next phase of the trials, he constructed a bigger chute out of a bedsheet and took his own leap off the backyard shed, breaking his arm and temporarily putting a hold on the program. Barely a year later, he fell again, this time from a tree stump while on a Sunday visit with family friends, and he broke his hip. It was a bad fracture and a subsequent infection resulted in osteomyelitis, a potentially debilitating bone disease. The Piantanidas were informed that their son would likely never be able to participate in competitive sports and certainly not contact sports. In addition, doctors confided, tests showed that the disease would in all likelihood leave the boy sterile. While recuperating at home, Nick, unperturbed, constructed a cardboard cockpit and instrument panel and invented an elaborate game in which he could fly combat airplanes from his sickbed.

Word got out in the neighborhood about Nick's condition, and everybody felt awful. He hobbled around on crutches for awhile. Then one day Jimmy Lagomarsino who lived in the next block up on Third Street caught a glimpse of Nick sprinting past his porch.

"Hey, Nick, are you crazy? What are you doing?"

Nick never even slowed down; he just shouted over his shoulder, "See you later, Jimmy!"

And that was it. By the time Nick reached high school, any infirmities were long forgotten. He'd grown tall and lanky and had shown himself to be a natural athlete. He'd learned to scuba-dive in the Hoboken YMCA pool and had taught himself karate. Sports were big in the neighborhood culture. But in spite of the fact that the world's first organized baseball game had been played just a couple of miles away in Hoboken and everybody played stickball in the streets from the time they were old enough to chew gum, the real passions in Union City were for football and—especially—basketball.

Nick practiced anywhere he could find a hoop and semilevel ground. There was a park about a block and a half from the house, and every summer morning the neighbors could watch him come dribbling down the stoop and up the Third Street sidewalk. He'd turn the corner, come to a jump stop, pivot, and let fly a set shot that would clear the bottom rung of the fire escape about fifteen feet up the side of an apartment building. He'd catch the ball before it hit the ground and then take off at full dribble for the playground. On winter weekends, the other neighborhood kids would wait until they heard the familiar racket that announced Nick was shoveling snow and ice off the basketball court before they'd bundle up and head to the park themselves.

Nick easily made the junior varsity Emerson High School basketball team as a sophomore, but abruptly quit when ordered by the coach to put out the cigarette he lit up in the locker room after a preseason game. It was an early manifestation of a trait that came to form the very core of Nick's character. He would be incapable of defining or even accepting success on another's terms. By temperament, his would always be a solo act.

But he was resourceful. He resolved to teach himself the game of basketball. He practiced alone. He studied books written by famous coaches. Nick's cousin and Hoboken hoops legend Nicholas "Nat" Hickey—born in Korcula as Nicola Zarnecich—had advised Nick to attend professional games at Madison Square Garden whenever he could and to copy the players' moves. Hickey, probably the first Croatian player to play in the National Basketball Association, had been an important figure in the early days of professional basketball, playing for the Original Celtics and the Cleveland Rosenblums in the 1920s, and for a handful of teams in the National Basketball League in the 1930s and early '40s. While coaching the Providence Steamrollers during the 1947–48 season, Hickey had appeared as a player in one game, giving him the distinction, at age forty-six, of being the oldest man ever to play in an NBA game. The pro game, Hickey counseled, featured a more aggressive, hard-nosed style of play than that on display in the highly regimented college programs. Nick took up his cousin's advice with a vengeance. Without the benefit of ever having played a minute of regulation high-school ball, he transformed himself into one of the premier young players in the New York City metropolitan area. People knew his name, and Nick would eventually become a valuable commodity in the semi-pro world, earning $50, $100, or even $250 for a single game.

As a teenager playing in the Union City Police Athletic League, Nick beat out future NBA great, Hall-of-Famer Tom Heinsohn for the league's Most Valuable Player trophy. In 1953, playing for a Hoboken YMCA squad in a celebrated national tournament that featured legend-in-the-making Wilt Chamberlain, Nick was the high scorer and shared YMCA All-America honors with Chamberlain. He later outplayed Lenny Wilkens, who would become not only a great player but the winningest coach in the history of the NBA, and was named Most Valuable Player at one of the renowned Mount Carmel tournaments.

Mount Carmel is a Catholic elementary school in Jersey City where an entrepreneur named Dominic Matticola once organized annual basketball tournaments that were New Jersey's precursor to Harlem's celebrated Rucker League showdowns. This was near-NBA-caliber ball, and only the toughest, flashiest New York City–area talent showed up. The gym was small, with short bleachers along one side and a stage at one end. In the basement, Matticola set up a full liquor bar for the paying customers.

Mount Carmel ball, like that played in the Rucker League, was freewheeling and entertaining for the players as well as the fans. The referees were there, and they were top-notch, but they mostly stayed out of the way and simply allowed the brutal ballet to happen. In spite of his lack of high-school experience, Nick had proved that he was for real and these tournaments became his personal showcase. "Nick always played to the crowd," says Jim Lagomarsino, who grew into a good journeyman ballplayer himself. "But he was only truly great against great competition." And that's what Matticola's tournament offered. Nick's on-court partner at Mount Carmel—he could usually be found filling the right-hand lane on the fast break—was another Union City stud named Eddie Madsen. Eddie was as skilled and as tough—rock-solid and street-smart and unforgiving—as Nick. Either of them was a threat to score fifty points whenever they stepped on the court. "They were flamboyant," one observer noted, "but they could back everything up. And they both insisted that the game be played the right way." In other words, woe unto the ball-handler who failed to hit the open man on a fast break. Nick once chased a ball-hogging point guard right out of the

Mount Carmel gym and onto the street. "When Eddie Madsen's open on the wing," Nick yelled with fire in his eyes, "he gets the ball! Every time!"

As time went on, Nick and Eddie formed a bond off the court. They hopped the train into Manhattan, chased girls, hung out at the Union City burlesque houses like the Transfer Station and the Hudson Burlesque, and—whenever and wherever necessary—defended their reputations with their fists.

Nick and Eddie had a pact: if one fought, both fought. It was automatic. "And we fought a lot," Madsen recalls. "But you never fought if you knew it was gonna be easy, see. That's wasting it. You only fight when there's a challenge. The worst thing that can happen? You lose. But you win inside." It was a code absorbed from the weathered bricks and sidewalks of Union City. "I never saw Nick in a fight that wasn't a tough one. No bullshit. And a challenge for Nick was just a problem to be solved. His trick was to dominate you mentally. That was his edge. And I never—*never*—saw anybody beat him."

Madsen was All-State in high-school and had forty-five scholarship offers waiting when he graduated at age sixteen. He made the starting squad at St. Bonaventure as a sophomore. But he had more than basketball and fighting in common with Nick. "I never had the discipline or commitment to get through college," he admits. Madsen was a scrapper, his specialty the brief flurry. He became a heavyweight boxing champ in the Army.

Unlike Nick, Ed reveled in the role of the big shot: the clothes, the jewelry, the sycophants, the loose women or, as Ed would say with a smile as big as the night, "Da whole nine yahds!" He fell into the fast life in the stylish clubs of Manhattan, worked briefly as a bouncer, then as a model, and later found work in movies playing tough guys, mobsters, and killers. He eventually bought his own bar in Union City. But after all that, drinking with Tony Bennett, working with Robert DeNiro, Lee Marvin, and Charles Bronson, Ed says, "Nick was the greatest guy I ever knew. Not even close."

And if Nick were around today? "Some business connected with speed: race cars, or maybe high-performance aircraft. Something with a technical dimension. He'd be in charge, but he'd also be down in the trenches alongside the guys doing the work. That's how Nick was."

Satisfaction came hard to Nick, and he would never cease the search for something beyond the moment. Jim Lagomarsino again: "He could be a difficult guy to spend time with, even for his friends. He was never what you'd call 'easy-going.' He was always pushing, stretching, challenging. He was never really comfortable in a group. And of course he'd disappear at the drop of a hat." Another member of Nick's Union City circle, Fred Cranwell, who became a journalist and sports information director for St. Peter's College, puts it another way: "Nick ad-libbed his life. He didn't believe in scripts. He believed in himself. He was the ultimate survivor. If a piranha bit Nick, the piranha would die."

Halfway through his junior year at St. Michael's High School (where he played football and scored a celebrated touchdown that ended a St. Michael's sixteen-game losing streak), Nick again dropped out, this time to start a scrap-iron collection business. The work toughened his body up in ways basketball couldn't, and it hardened his resolve. In spite of a lifelong two-pack-a-day cigarette habit, Nick—like his father and his brother—would always keep himself in top physical condition. He was a runner and a swimmer, but never a weightlifter. Nick tended to focus on the specific skills he would need for the activities in which he wanted to participate.

At his parents' request, he returned to finish school before joining the National Guard. Not much later, he volunteered for the Army and was assigned to two years of stateside duty which he served at Camp Polk, Louisiana, and Fort Riley, Kansas. He was All-Army in basketball both years and, like Ed Madsen, took up boxing—with similarly impressive results. He fought eleven times as a heavyweight and left the service undefeated with the rank of corporal in 1954. "I almost made sergeant once," he would say, "but they busted me before the papers could get through."

In 1955 Nick entered Fairleigh Dickinson University on a basketball scholarship in head coach Dick Holub's program. It looked like the perfect situation. Nick immediately led the FDU freshman squad in a preseason scrimmage rout of the varsity. Then, on November 7, Nick went solo and abruptly quit the team under mysterious circumstances. Holub, whose big plans were already being built around his freshman sensation, was furious and reportedly vowed that Nick Piantanida would never again suit up for any college in the United States of America.

Angel Falls

While serving his two years of active duty with the Army, Nick spent a lot of his down-time thinking about getting out. Not surprisingly, he chafed at the discipline and the rote learning, just as he had in the classroom and the locker room. One night, killing time in the barracks, Nick came across an article on diamond hunting in a men's adventure magazine. The article described the interior of Venezuela as a nearly impenetrable jungle where a fortune in gold and precious stones was available for the taking. It led an already romantic imagination to Sir Arthur Conan Doyle's *Lost World* and to fantasies of El Dorado deep in the heart of South America's Gran Sabana. Nick conjured up an expedition to the valley of the Rio Churun where the tumbling waters of Angel Falls formed the highest waterfall in the world: 3,212 feet, fifteen times the height of Niagara Falls. It was one of the Seven Natural Wonders of the World, and Nick announced that he would set out to be the first man to climb the cliffs of the Auyán-tepuí, or Devil Mountain—technically a giant mesa—from which the falls erupt. (The legendary aviator Jimmy Angel had crash-landed a small airplane, a Metal Flamingo, on the treacherous mesa top in 1937. Angel lived to tell the story but never actually reached his eponymous falls.) Nick, with few financial resources, decided that he would manufacture publicity and locate sponsors to bankroll the trip. And once in South America, he'd collect his fortune in diamonds and return home a wealthy man. Most of his Army buddies assumed it was just another of Nick's wild schemes.

But once back in New Jersey after his discharge, Nick convinced one of his Army friends, Walter Tomashoff, another outstanding athlete who had played football for New York University, to join him in Venezuela. True to his word, Nick began to sign up corporate sponsors intrigued by the idea of an association with the first man to scale the cliffs of Devil Mountain. No one, it seemed, could say no to the Piantanida charm and confidence. In amazingly short order, Nick had wangled an outboard motor from Evinrude, firearms from Colt, cameras and film from Kodak, and passage for two to Caracas from Grace Lines Shipping.

In the meantime, Nick and Walt set about training themselves in the specialized techniques of mountaineering. Devil Mountain promised to be an extremely challenging technical ascent, assuming the climbers could even make it through the dense jungle to the base of the massive structure. Yet Nick had somehow neglected to mention to his corporate benefactors that neither he nor Walt had any experience whatsoever with climbing ropes,

Nick (with climbing rope) and Walt Tomashoff board the SS *Santa Paula* bound for Venezuela. (Courtesy Vern Piantanida)

pitons, or carabiners. So, using the Hudson River Palisades above Hoboken (the cliffs from which the original stones had been quarried to pave the streets of Manhattan), some gear purchased from a Jersey City sporting goods store, and a couple of illustrated books from the library, Nick and Walt went to work to master the required skills. They climbed by day and supported themselves with night jobs at Continental Can in Paterson. And while Nick had put competitive basketball on hold while he prepared for Angel Falls, he did agree—at the urging of his supervisors—to participate in a national free-throw shooting contest. Companies all over the country were invited to enter the contest, and Nick represented Continental Can. Each shooter was given fifty shots from the free-throw line. Nick missed his first attempt and followed it with forty-nine consecutive makes, easily securing the first-place trophy.

Nick made good use of his time on the six-day voyage to South America in late October 1956, befriending the captain of the SS *Santa Paula*, who agreed to arrange an introduction to the former editor of the *Caracas Daily Journal*, Ruth Robertson. Robertson, the first to calculate the actual height of Angel Falls, had made a landmark expedition to the falls' base in 1949 which she had chronicled in a long article for *National Geographic* magazine. On their last full day on board, Nick won a cigarette lighter for placing first in the ship's diving contest and promptly gave away his award to the prettiest girl he could find. Ilse Sobernheim, who was traveling to her Caracas home with her family, was flattered and insisted that Nick visit her in the Venezuelan capital before embarking on the jungle expedition.

Once in Venezuela, Nick wasted no time in looking up Ruth Robertson. She agreed to help Nick and Walt and put them in touch with Alejandro (Alec) Laime, an adventurer with unparalleled knowledge of the Guiana Highlands region who had guided Robertson's trip. Unfortunately, Laime revealed that he had successfully climbed Devil Mountain with another National Geographic Society group only a few months earlier. So, after a quick study of maps and some agitated discussion, Nick revised his plan: He and Walt would set out instead to become the first to scale the more difficult, some said unclimbable, north (or wet) side of Devil Mountain, literally alongside the great falls. A skeptical and somewhat testy Alec Laime agreed to accompany them, but only after they agreed to listen to his stories of some of the earlier failed expeditions into what Robertson's *National Geographic* article had called "this weirdly beautiful high jungle of impenetrable . . . mesas like mighty fortresses a mile to two miles high." The scale of the venture was truly staggering: the great sandstone massif that forms Devil Mountain is twenty miles long and twelve miles wide. In one attempt to reach the base of the falls years before, an entire boatload of twenty men had simply disappeared. It would take an extraordinary effort simply to reach the cliff face and a nearly superhuman one to make it up the wall to the top.

It would be a wait of nearly two weeks before the expedition could leave Caracas, much of that time spent haggling with suspicious local officials over permits that would be needed for the pistols Nick and Walt had brought with them. (This exercise in patience and persuasion would serve Nick well years later in his encounters with more formidable bureaucracies as he would attempt to organize and outfit himself for a trip into an even more exotic region: the extreme skies of the stratosphere.) During the delay, Nick looked up some of the local newspapermen. Photos of the American climbers appeared in several Caracas publications and one paper ran a lengthy interview with Nick.

On one of their last nights in Caracas, Nick and Walt dined at the Sobernheim house where they listened to records by the new American rock 'n' roll sensation, Elvis Presley, and later went to the theater on a double date; Nick had asked Ilse to find a girl for Walt. "On the way to Ilse's house," Nick wrote in his journal, "I kissed her and the cab driver promptly pulled over. He instructed her, in Spanish, not to do that because it is against the law." Nick added, "Some of these Venezuelan customs are very annoying, especially to an American."

Three weeks later, after setting out from Canaima on a punishing trip by dugout canoe up the forty-eight sets of wild rapids of the Rio Carrao and the Rio Churun, a journey filled with minor adventure and major hardship (rain, mud, biting insects, giant spiders, infected wounds, and blistering skin), Nick, Walt, and Alec reached the base of Angel Falls. The trio made numerous probing approaches with heavy packs, hacking nearly vertical paths through the prickly, snake-infested brush, trying to find a climbable route. In places the jungle was so clotted with vegetation that clearing a one-hundred-yard trail required ten to twelve hours of hard labor. Ironically, the cliffs were so thick with roots and small trees that the climbers were never even able to use the ropes and pitons that Nick and Walt had worked so hard to master. Nick's cheap boots literally rotted away and he was forced to continue in his Chuck Taylor All-Star basketball sneakers, which eventually split apart themselves and had to be reinforced with loops of cord.

As the 100-degree days dragged on and their food supply diminished, the men began to wear on each others' nerves. Nick, in spite of a broken thumb and badly injured feet, became frustrated when his companions were unable to match his energy and determination. On December 5 he made the following entry in his journal:

> It began raining last night about 9 P.M. and did not let up until 10 A.M. this morning. It continued to rain on and off all day. The river has risen from 3½ feet to four feet since yesterday and the water is whipping along. Because of the rain, Walt and Alec decided to stay in camp today as any normal person would. Not being normal, I left camp at 7:30 A.M. with my machete and lunch and headed out in the direction we were going.

On December 12 Alec Laime gave up in disgust and announced his intention to return to Caracas. A confrontation ensued. Alec and Nick argued about Laime's hammocks and blankets, which the guide intended to take back with him. Nick agreed to return the equipment to Laime, but only after they'd summitted Devil Mountain. Following some bitter words, Laime abandoned Nick and Walt and made his own way back to civilization.

On December 19, 1956, ten days after the waterfall's namesake Jimmy Angel died in a Panama hospital from injuries suffered in a small plane crash that summer, the journal records triumph:

> It rained again last night and the jungle is wetter and more slimy than usual. We reached yesterday's stopping point. The going from here was very bad. The jungle would thin out at intervals only to become thicker than before. Heavy clouds were hanging in the canyon and that made the situation worse. We hoped they would lift so that we would get clear pictures when we reach the top. Upon reaching a point where the canyon leveled off, I climbed a tree to determine where we were. The canyon took a sharp turn to the right about 100 yards in front of us. I was unable to determine whether it ended there or led to the top. The only thing to do was to keep on going. We reached the turn about 11 A.M. and found a sharp incline leading up to the crest. We became excited as we cut through those last few yards. It was still cloudy but cleared sufficiently so that we could take some pictures. We had made it to the top—conquered the north side of Devil Mountain. After a short stay, we had to hustle down because we did not have any food left. We all but ran down the mountainside.

The pair spent most of Christmas relaxing in Laime's hammocks back in the

Indian village of Canaima, sleeping off hangovers and a feast of ducks that Nick shot with a borrowed shotgun.

Upon returning to the States in February, Nick found himself being interviewed on the *Today Show* as a conquering hero. But neither Nick nor Walt ever saw any diamonds or anything else of much material value in Venezuela. They returned to their lives in New Jersey as penniless as when they'd left.

Nick promptly found a Catholic Men's League basketball team that needed an extra player and scored sixty-three points in his first game. Along with old Third Street cohort Jim Lagomarsino, Nick led the team to the New Jersey state championship.

Over drinks one night in Caracas, Ruth Robertson had regaled Nick and Walt with stories of an unnamed and unmeasured waterfall she'd once spotted from a bush plane while surveying the largely mapless jungle beyond Devil Mountain. These falls, in a region frequented by Indian groups more hostile than the gentle Camaratas Nick had encountered on his previous trip, would present even greater challenges. But Robertson had judged these falls to be even higher than Angel Falls, and so Nick, while working odd jobs and playing basketball, made plans for a second expedition to Venezuela in December of 1958. It was a good excuse, anyway, to escape the mundane world and set out on another adventure: "I decided nothing would stop me. Glorious thoughts of fame as the discoverer of a yet higher falls filled my mind."

After three days of strategizing in Canaima, Nick headed once again up the muddy, iron-stained froth of the Rio Churun, this time without Walt Tomashoff or Alec Laime but in the company of an affable Dutchman named Rudy Trufino whom he'd met on his previous visit, and Trufino's Dalmatian pup. The pair encountered tapirs, brilliantly colored parrots, monkeys, wild pigs, and, on the fourth day, a black panther swimming the river. It was slow going due to low water levels. It was the driest season in fifteen years, requiring them to haul Trufino's boat through long stretches by hand. The daytime temperatures were in the mid-90s and Nick and Rudy were plagued by swarms of biting wood ants, tarantula-infested campsites, and packs of oversized jungle rats after their food supplies.

In spite of the annoyances, the trip wasn't total misery. Nick found time to record some almost bucolic moments in his journal in the peculiar pur-

Nick muscles Rudy Trufino's boat through a rapids on the Rio Churun. (Courtesy Janice Post)

plish prose style of the adventure books he'd been reading: "Mating calls of a countless number and variety of birds filled the air with a majestic euphonious accord, disturbed only by the occasional throaty screech of the king parrot."

A week into the trip, they passed around the bulk of Devil Mountain and continued deeper into the bush. But when, late one evening, they reached the location described to them by Robertson, they were devastated.

"I strained my ear to the ground," Nick wrote, "trying to hear the rush of water from the new falls. I could only hear the distant Angel Falls. I waited impatiently for morning. By 11 A.M. the remainder of the canyon was before me and I was disappointed. All we found was a dried up remnant of a falls. . . . Rudy had pain in his eyes and asked how high I thought this dry falls was." Nick estimated them to be 2,700 feet high at best, inferior to Angel Falls by several hundred feet.

On their dejected return to Canaima, Nick and Rudy began making notes for a brochure to advertise a guide service to Angel Falls. They reasoned that they could lead two expeditions of tourists a season at a round-trip cost from Miami of $1,250 per person. If they could organize groups of fifteen paying customers a trip, they could easily clear more than $10,000 apiece per season. It paled in comparison with the original fantasy of diamonds and gold, but for Nick at least it beat a factory job in Union City. A mock-up of the brochure introduced "Your Guides":

> "Jungle" Rudy Trufino lives in the jungle near Angel Falls with his wife and child. . . . Prior to coming here, he lived in Holland, attended several schools in Europe, and is a graduate veterinarian. On several occasions, Rudy performed emergency operations on persons requiring immediate attention; and doctors back in city hospitals commented favorably on his work.
>
> Your other guide and leader of this expedition is Nick Piantanida, a native of New Jersey. Nick, 28, has also spent considerable time in the Gran Sabanna. In 1956–57, he led a mountain climbing expedition which became the first to successfully scale the north face of Devil Mountain. . . .
>
> He is a competent pilot, skin diver, and boatsman.
>
> It can be said without any hesitance, that a more competent pair would be difficult to find.

While Trufino would go on to found a successful guide service known as Jungle Rudy's—a service that would eventually be run by his daughter in the jungle near Angel Falls—for Nick it would be nothing more than another pipe dream. Upon reaching Canaima and saying his goodbyes, Nick boarded a plane and returned broke and weary to winter in New Jersey. On the flight out, Nick convinced the pilot to make a pass over Auyán-tepuí and got his first look at Jimmy Angel's airplane, still resting on the broken mesa top where the explorer, his wife, and his partner had abandoned it more than twenty years before. The plane, which Angel called *El Rio Caroni*, was recovered by helicopter in 1970 and can be seen today outside the airport terminal in Ciudad Bolivar. A replica has been placed at the original crash site on Devil Mountain. Angel's own ashes were scattered over the falls that bear his name in 1960.

Nick told a Union City reporter that the failure to discover what he had hoped would be the world's new highest waterfall would not deter him from future adventures. "You can't tell what moves you to do such things," he explained. "That is just how I am. Most people talk about such things and do nothing. I just have to go and see."

Nick would take work briefly, and surely reluctantly, as a supervisor in a

Union City embroidery factory. He would have to wait several frustrating years before finding another challenge as compelling as Angel Falls.

Dodge City

In September of 1959, with few other prospects, Nick followed his brother to Dodge City in western Kansas. Vern was in his final year at St. Mary of the Plains College where he had become something of a phenomenon in his own right: a dean's list student and scholarship football player—captain of the team, in fact—who had been elected president of the senior class. The nuns who ran the college had all fallen in love with their East Coast prize and encouraged him to recruit other New Jerseyites who could make contributions to the college, especially if they could help shore up the school's athletic teams. Vern was an effective ambassador. He had adjusted easily to peaceful Dodge City and had already met Paula, the local girl who would become his wife. In short order, Vern helped secure football scholarships for a few friends from the old neighborhood and—with a bit of trepidation—convinced the nuns to offer a basketball scholarship to his itinerant older brother.

Vern arranged for Bob Taglieri, Marty Familetti, and Nick to share an apartment near campus. They would later be joined in the apartment by another talented, free-spirited ballplayer when Vern's high-school friend Jack Nies, who would go on to enjoy a long career as an official for the National Basketball Association, transferred from Georgetown. Bob and Marty arrived first. Nick didn't show up until the day before classes were to start, but he hit Dodge like a mythic gunslinger swept in off the old Santa Fe Trail. Bob and Marty knew Nick mostly by reputation and were uncertain of just what to expect. They, of course, knew all about Nick's basketball prowess, and they'd heard about Angel Falls. Bob recalls his first encounter: "This monster, this god, walked through the door. From that moment on, Marty and I were both terrified. Nick had these long arms and huge hands, like a gorilla. And hairy. Nick had hair on his fingernails."

Nick brought with him a parrot, some snakes and lizards in makeshift terrariums, and a small alligator. The alligator, Nick announced, would live in the bathtub.

"But what if I need to take a shower?" Bob asked helplessly. Nick shrugged. Bob and Marty took a lot of showers in a neighboring apartment that year.

In short order, Nick had cut a wide and colorful swath through Dodge City. On his first night, he reportedly seduced the most beautiful woman in town. Over the course of the year, while Bob and Marty were daydreaming—mostly in vain—of simply getting a date with an actual girl, Nick became notorious for a dizzying string of liaisons with prominent upperclasswomen and local waitresses. His favorite rendezvous spot was the town cemetery.

But nowhere did he stand out like he did on the basketball court. He was a left-handed player, but he liked to warm up shooting right-handed jump shots in order to hide the extent of his skills before a game began. As a freshman (albeit a twenty-seven-year-old freshman), he led the St. Mary's Cavaliers, the league, and the conference in scoring, averaging 27.3 points per game and making the Catholic College All-America list by season's end. Nick was an intimidating player with a flashy New York style that dazzled opponents and spectators alike. Although he stood only 6-foot-1, he could dunk left- or right-handed and was typically assigned to guard the opposing team's best player, regardless of size.

Once, at an away game versus the Tabor College Bluejays in Hillsboro, Kansas, Nick had been struggling to guard a much taller player in the low post and was pulling out all the tricks he knew, one of which involved holding the opponent's leg with his off-hand away from the referee's view. And while the referee could not see what was going on, the crowd—which definitely could—began to heckle Nick. When the game ended and the teams filed back into the locker room, Nick remained at center court and engaged the crowd of big Kansas farmers with a cocky grin. As the booing grew louder, Nick challenged anyone who was interested in testing him to step onto the court. There were no takers.

Nick was not only a star athlete and a member of the Plains Players, the school drama club, but a good student who excelled in the classroom, seemingly without ever studying. He liked science, especially biology, but found nothing in the St. Mary's curriculum that excited him. Nick wrote well, however, and quickly landed the job of sports information director for the school. In that capacity, and writing under a pseudonym, Nick authored numerous press releases on his own exploits as a basketball player. He also made a personal appearance at an American Legion function in Scott City, Kansas, at which he narrated a slide show of his Venezuela trips and offered to guide any interested parties to Angel Falls the following summer for a price of $1,000 per person. It was the first of many such appearances he would make over the coming years. In addition to his photographs, he would display for his audiences a collection of artifacts that included authentic curare-tipped blow-gun darts, the pair of lacerated sneakers that had made it to the top of Devil Mountain, and a spear point that he'd gotten from a Camarata hunter and which he kept in the trunk of his car for the purpose of opening oil cans. Nick liked to end his Venezuela lectures with a defense of risk-fraught adventure by listing important discoveries made by early explorers in South America. His standard closing line: "Many people are itching for an opportunity. The problem is, they're itching instead of scratching."

And, as so often happens, Nick's successes on the basketball court resulted in special treatment—notably from the college administration. His outright refusal to wear the traditional St. Mary of the Plains freshman beanie—traditionally required of all first-year students—resulted in an ad hoc exception to the rule for any veterans of the armed forces. And when one of the nuns balked at giving Nick a key to the gymnasium so that he could practice his jump shot whenever the desire moved him, he locked her in the gym for the better part of an evening, which resulted in a new policy that kept athletic facilities open at any hour upon request.

Nick also forged a reputation beyond the confines of the campus. He would take Bob and Marty to a local laundromat to "ride" the dryers. (They would switch off the heat, climb inside, and go for a spin.) And of course Nick fought. Ethnic-looking Catholic boys with New Jersey accents were fat targets in western Kansas in the late fifties, and Nick seemed to relish the opportunity to antagonize the locals and—if necessary—to put them in their place. None of his Dodge City buddies can remember Nick ever losing a fight.

The voluble Bob Taglieri describes what it was like to run around with Nick Piantanida: "If Nick said, 'Come on,' you went. You never knew where. But things always happened when you were with Nick. He was fearless. Absolutely fearless." Bob went on to become a two-time Catholic College All-American on the football field and, later, a successful head coach with a knack for transforming weak programs into powerhouses. But he never again met anyone who affected him the way Nick did. "Nick and my father were the only two real *men* that I've known," he says. "Nick was what you wish you had the guts and brains to be."

During the second semester, Nick finally met his match when he fell hard and uncharacteristically in love with a well-mannered, attractive young woman from Kingman, Kansas, named Pamela Stewart. The relationship quickly turned serious. Pam's father was a judge and made no secret of his opposition to the brash Yankee eight years his daughter's senior. But it was Pam's mother who finally broke up the match. Nick urged Pam to defy her parents, but she was unwilling.

Nick—a magnet for the opposite sex since his teenage years and largely unfamiliar with rejection—was openly heartbroken. It was the first time any of Nick's buddies had seen him cry. He was angry at Pam's parents, but angrier at her for what he could only feel was a betrayal. It may have been the first time in his life that Nick had gone all-out after something he wanted badly—and lost. Bob Taglieri believes it was a watershed event in Nick's life. He's convinced that it was the breakup with Pam that drove Nick away from St. Mary of the Plains at the end of the school year, never to return. Afterwards, Nick was a little more subdued and noticeably darker in his outlook. Bob even thinks that the experience somehow drove Nick to search for some ultimate challenge, some impossible dream.

The Kansas/Jersey connection, established by Vern Piantanida and bolstered briefly by Nick's presence, would be in effect for several more years as Vern accepted the jobs of assistant professor of mathematics, athletic director, and basketball coach for the college and continued to recruit promising athletes from the East Coast. Nick, meanwhile, reverted to life as a fly-by-nighter attracted by whim to grandiose plans and get-rich-quick schemes. He became briefly fascinated with gambling. He had never been more than a mediocre poker player, at least according to his Union City cohorts, but he became convinced that a strategy of doubling up on losses—an idea responsible for the separation of a lot of otherwise intelligent gamblers from their money over the years—was the formula for beating casino games like roulette and craps. He found somebody to loan him $1,000 and took off for Las Vegas, returning broke a few days later. He then turned his attention to horseracing, with similar results. He fared only slightly better as an entrepreneur. He got involved with a product called Traction Tracks for vehicles stuck in snow or ice. He sold vitamins. And he came close to investing his entire life savings, which in truth didn't amount to much, in a scheme to raise nutria for pelts to supply the ersatz-mink-coat market.

Nick's attraction to the creative scam was completely natural. Skill at the scam, the hustle, was a badge of honor on the streets of Union City. It was a tactic for protecting dignity—not a criminal thing, but a survival skill for those without financial resources, connections, or a formal education. It was a way of evening the score with the world and, for some, it was a shortcut to respect and respectability. Unlike Vern, who would choose the more laborious and prosaic overland route, Nick always had a keen eye out for the shortcut.

It was a puzzle to Vern why his mercurial brother, in Vern's eyes a more gifted human being, could not do as he did and apply himself academically. Both of the Piantanida boys were natural leaders, but Nick could never abide the stasis demanded by high school, the Army, and now college. It certainly wasn't from a shortage of brains. Their parents always insisted that Nick and Vern had scored identically on IQ tests. But Vern, the math major, never believed that, always insisting that Nick was smarter.

"But what did Nick do with his gifts?" It was a question Vern would be asking decades later, with a deeper resonance.

In the fall of 1960, Nick showed up on the campus of the University of Tampa where he'd been offered yet another basketball scholarship. But, almost predictably, within a couple of months he was gone. There were rumors that he'd thrown the coach against the locker-room wall. Years earlier he'd received some inquiries from the venerable University of Kansas coach Forrest "Phog" Allen, but hadn't relished the prospect of feeding the ball to Kansas phenom Wilt Chamberlain. Nick had always wanted the ball himself. The solo act required it. Once again, he decided to join his studious little brother who was now enrolled in graduate school at Wichita University (later to become Wichita State). Wichita had its own high-powered coach, future great Ralph Miller. Miller offered Nick a scholarship based on some casual workouts with the team, but with his NCAA eligibility expired and not wanting to jeopardize the Shockers' season or to compromise the reputation of Miller, whom he liked, Nick did the right thing and walked away from the team.

While finishing out the semester—Nick made an issue of a loophole he'd discovered in the university's requirements policy that allowed him to avoid the general chemistry course required of all biology majors, an effort that later caused the school to rewrite and clarify its policy—he began to take a renewed interest in animals and animal husbandry, keeping a number of specimens in his dorm room: some two hundred ground squirrels, snakes, baby alligators, mynah birds, even a monkey. When the semester ended, Nick left school—for good this time—and drove back to New Jersey in a stationwagon full of reptiles and rodents and a head full of plans.

On the trip home, the car radio might have picked up the breaking story of Russian cosmonaut Yuri Gagarin who on April 12 had become the first man to orbit the earth.

Janice

Nick opened his own business on Summit Avenue in Union City late that fall: Animal Wholesalers, Inc. The plan was to stock the company's inventory with specimens collected on various expeditions to Central and South America, to establish relationships with other animal distributors around the world, and to provide exotic creatures to zoos, circuses, and wealthy private collectors. It dovetailed nicely with another plan Nick was working on that involved filming a series of documentary television programs in Central and South American locales.

When Nick acquired his first cobra, a mail-order item from India—cost: $6.50 FOB, Calcutta—he took it with him to the Ichi-Ban. This was a tavern owned by an amiable Union City hellraiser named Bobby Santomenna who'd once lived in Japan, and it may have been the most unusual bar in all of Hudson County—which is saying something since Hoboken alone was famous for being home to more bars per square foot than any city in America. For decades, bars had been important social institutions for the region's immigrant culture, serving as secular meeting halls where those from diverse backgrounds and walks of life could mingle and ideas could cross-pollinate.

The Ichi-Ban was New Jersey's self-contained Greenwich Village, frequented by hipsters, athletes, musicians, bookies, prostitutes, and a certain breed of curious celebrities. From the ceiling hung medieval weapons, women's underwear, a traffic light, stuffed Halloween costumes, a wheelchair, a plaster cast from somebody's arm, cages full of live birds, and taxidermy of all description. In spite of the fact that there was a boxing ring in the center of the floor complete with turnbuckles and ropes, fights were rare in the Ichi-Ban. Santomenna—a sweetheart of a guy as beloved by the offspring of the Union City immigrants as he was held in contempt by the religious leaders and pious politicians—regularly regaled the patrons from a microphone he kept behind the bar. He liked to announce specific services that might be available from the prostitutes in attendance and was not averse to holding auctions on their behalf when business was slow.

Nick was a regular. And Bobby always lit up when Nick walked in. "He was the kind of guy I wanted in the place," he says. "He always brought some kind of fun in with him." So when Nick walked in with the cobra in a pillowcase, everybody gathered around. Nick had announced his intention to learn to milk the venom from the cobra, but first he had to teach himself how to handle it. The crowd in the Ichi-Ban offered ideas. Finally, Bobby hit on the solution.

"Nick," he said, "we'll just sew the little bastard's mouth shut, see? You can practice a little, and once you've got it down, we'll take the stitches out."

That's precisely what they did. Aided, no doubt, by whatever Bobby happened to be pouring that day. Nick learned to handle the snake, but at some point—who knows what distractions might have arisen in that place?—they lost track of the cobra and it disappeared. Word got out, and for days a good portion of Union City was scouring its closets, basements, rooftops, and alleys looking for Nick's cobra.

Nick got more cobras and before long was staging battles-of-the-species in a plywood-fenced arena he and Bobby built in Nick's parents' backyard: cobra versus mongoose, cobra versus goanna lizard; cobra versus Nick. Crowds gathered and even local cops were stopping by to watch the action. It's safe to assume that wagers were placed.

Then one afternoon in February there was a dance contest at the Ichi-Ban, and Nick found himself intrigued by the girl who won first prize: a pretty, trim Irish brunette he'd never seen before. He learned that her name was Janice McDowell and that she lived just seven blocks away in North Bergen.

"I rarely dated and I never went to bars," Janice explains. "This was unusual, this dance contest. And I left immediately after it was over."

Whether he was still hurting from the romantic snub in Dodge City or not, he found himself once again transfixed. Janice had graduated from high school in the spring and was working now at DuraTest in Union City. Nick asked if he could walk her to work, but she was put off by his cockiness. She'd seen enough street attitude to last a lifetime, and she wasn't impressed.

"Of course, Nick knew all the cops," she says, "and every time I'd pass one on my walk to work, they'd start lobbying for Nick. 'Come on, Janice,' they'd say. 'Nick's a nice guy! Why don't you give the poor guy a break, huh?'"

Nick began waiting for her each morning and driving at walking speed alongside her, blocking traffic and creating a commotion she found both flattering and embarrassing. Eventually she relented and agreed to let Nick chauffeur her to and from work.

"He was the most handsome man I'd seen in my entire life," she admitted later. But at first, she couldn't understand what he saw in her. "I was plain Jane. None of my friends thought it made any sense. Nick had girls following him wherever he went, and they all thought Nick should have been interested in them."

One day Nick showed her a scrapbook full of newspaper articles on his basketball exploits and heroic coverage of Angel Falls. She was quick to let him know that none of it impressed her. On the other hand, Janice's parents were crazy about Nick despite the nine-year difference in age. So unlike the parental sabotage he had faced in Dodge City, he found firm allies in the McDowells.

"From the day that we met," Janice says, "we rarely spent an entire day apart. You never would have seen any two people any more in love."

Rather than any achievement or talent, it was what Janice saw as Nick's reservoir of compassion that really got her. She claims he was incapable of ignoring anyone who asked for help. When his friends were in trouble, Nick was always the first one there to bail them out of jail or get them cleaned up after a fight. He emptied his pockets of cigarettes and coins for panhandlers. On their frequent trips into New York City, Nick would take down-and-outers from the Bowery to a diner and buy them meals.

But shortly after that first meeting, on a return visit to the Ichi-Ban, Janice got a look at another side of Nick Piantanida. A drunk made a pass at Janice and some words were exchanged. A few minutes later Nick and the drunk agreed to step outside and settle the matter. Janice moved to follow them when Bobby reached from behind the bar and put a hand on her shoulder.

"Don't," he said. "You don't want to see this."

No more than two minutes later, Nick came back in alone as if he were simply returning from a conversation on the other side of the room. "Call an ambulance, Bobby," he said.

She couldn't get Nick to talk about it, but Janice found out later that he had nearly killed the drunk. There was a latent rage in Nick that everyone agrees was there, but which few people ever saw. To counter the fiery temper, according to those who knew him, Nick possessed extraordinary self-control. Janice claims never to have seen Nick commit a violent act. Nevertheless, control aside, Nick *was* capable of serious lapses in judgment. He was once suspended from high school for peppering a teacher's car with a borrowed pellet pistol.

Such lapses, however, were early and rare. Nick was a genuinely religious young man—if not always a regular church-goer—and his reputation for honesty and integrity appears, in the main, to have been well deserved.

That summer Nick and Janice were married in a small ceremony at St. Joseph's Church in Union City. Vern was the best man. A few weeks later Janice was pregnant. This was a welcome surprise since Nick's childhood doctors had counseled the family to assume that he would never have children. As if to spite medical science, Nick fathered three daughters in quick succession: Donna, Diane, and Debbie. Over the years, those who knew Nick and Janice would describe the marriage as a solid and happy one, though the relationship never ceased to puzzle some of Nick's buddies. Nick prized Janice's beauty and poise and was as devoted to her as she was to him. Yet their personalities were fundamentally different, and while she supported his overweening ambitions to the end, she personally shared none of his desire to challenge and conquer. Later, on location for Nick's great parachute project, Janice could often be seen standing literally in the shadows, with seemingly no desire to bask in the glory that had become Nick's lifeblood, and often appeared to be on the verge of shattering under the relentless pressure and drama. Reporters who would routinely allude to her physical charms—it would be a rare newspaper or magazine piece that would not pair a reference to Janice with one of the adjectives "attractive," "pretty," "shapely," or "beautiful"—would seldom ask her for anything more substantive than her "feelings" about the project. Her typical response would be: "I trust in Nicky's judgment." As one observer said, "She was in the loop but not *really* in the loop. A lot of race-car drivers' wives, they're in there with their stopwatches in the pits saying, 'Go, Johnny, go!' She was more retiring, kind of an anxious presence on the sidelines. A real fifties chick."

Meanwhile, Nick and Janice had moved into their own apartment on 15th Street. She worked for a short time in a garment factory embroidering designs on towels while he concentrated on the exotic animal business. Nick purchased a female Royal Bengal tiger cub from the Impex Export Company in Calcutta for $850 and named her Natasha. He loved to cruise the streets of Hudson County with the tiger in the back of his stationwagon and watch peoples' reactions. The cub lived with them in the apartment for nearly ten months until she was large enough to put her paws on Nick's shoulders and tower a full head above his, at which point Nick arranged to trade Natasha to a group of filmmakers for a mechanically suspect twin-engine airplane.

In late 1962 Nick began to expand his ambitions as an animal dealer. He was developing a reputation among a number of monied clients in New York City for being able to deliver almost any kind of animal from anywhere in the world. He made trips to Florida to learn the finer points of handling and milking venomous reptiles from a noted herpetologist. He even did a modest business in snake venom and antivenins. Then, in November, he made a two-month trip to Mexico's Yucatan Peninsula in search of more exotic species. His plan had been to establish an Animal Wholesalers satellite office in Merida and to build a reliable supply pipeline from Central America to New Jersey. Nick wrote Janice long letters every other day and kept her apprised of his progress. Janice was due to give birth to Donna on Christmas Day, and Nick was increasingly nervous that he would miss the birth. "If it's a boy," he wrote her, "and I'm not home yet, you can name him anything but Nicholas." Then one day, without warning, he walked into the apartment with arms full of presents and a promise to curtail the extended expeditions. At least for a while.

Nick, of course, wouldn't learn about it until later, but while he was chasing snakes in Mexico that autumn, a Russian parachutist named Andreev had set a new world free-fall record.

Nick continued to play basketball whenever he could find a worthwhile game. According to Vern, it was during this period that Nick received an offer to sign with the NBA's New York Knicks for $5,000. Nick reportedly declined, claiming he could make more money barnstorming. In fact, even before the Knicks' offer, Nick had made the starting squad of the New Jersey Titans, a professional team slated for adoption by the NBA. But after several Titans players were implicated in a college point-shaving scandal, NBA Commissioner Maurice Podoloff disavowed any association with the team.

Vern also recalls that Nick was approached by promoters at Madison Square Garden who were looking for a "great white hope" to fight the flamboyant new Olympic gold-medalist and heavyweight champion Cassius Clay—and that Nick dismissed the idea out of hand. If true, the incident would seem to testify not only to Nick's innate good sense but to a practical awareness that he was not in fact invincible.

Brick Town

Nick's parents had acquired a summer home in Toms River on the Jersey shore. It had always been one of Nick's favorite places to go to unwind, to swim, to think about the future, and he was gratified to find that Janice felt the same way. By the early sixties, Union City was in transition. The burlesque houses had been closed by the Catholic city fathers and many of the first generation of Nick's and Janice's acquaintances had begun to move on. In a few more years, even Vern, having landed a lucrative job with IBM, and Paula would elect to move up the Hudson River Valley to Kingston to start a family of their own. Some of the European immigrant charm was being crowded out by new, unfamiliar cultures, and the shore was a breath of fresh air—literally. Nick and Janice would head down whenever they could escape the city for a few hours. They took long strolls on the boardwalk at Point Pleasant in the summer of 1963. Janice loved lying out on the beach on warm nights and listening to Nick talk his way through his various schemes and grand ambitions. And there was a recurring theme: Nick needed to find something that was different, distinctive, something that would be all his own, some way to make his mark on the world.

His favorite times were when the big storms rolled in on the jetty. He loved the power and the mystery of the sea and would fantasize about making the world's deepest dive in a bathyscaphe, carefully explaining the difficulties a super-deep dive would entail. It was one of the things Janice valued most about Nick, that he had the patience and consideration to make sure she understood everything he wanted to do and exactly how he intended to do it.

They slowly began to sever most of their social relationships and spent time almost exclusively with each other. Nick and Janice didn't go in for a lot of the typical entertainment opportunities of the day: movies, plays, nightclubs. As Janice says: "When you were with Nick you didn't really need entertainment. We were our own entertainment." But she does recall them parking near the beach for hours and listening to the radio. Nick's favorite song was an Acker Bilk hit called "Stranger on the Shore," a lush clarinet-and-strings number that had the distinction of being the first record to hit number one in the United States and England simultaneously. Nick told Janice that the song would always be dedicated to her. It was their song.

It was on a return drive from a particularly relaxing visit to Toms River that they had by chance passed the little Lakewood Airport and seen the parachutists falling out of the sky. "I need to learn to do that," Nick commented. He'd gotten his pilot's ratings while still a teenager, and he felt that—with some of the questionable aircraft he flew, especially the old World War II–era Cessna T-50 Bamboo Bomber he'd obtained in exchange for the tiger—that he really should acquire the skills to bail out should the circumstances ever require it.

It was the very next day that he drove back down to Lakewood and made that fateful first jump—the one that changed everything.

As Nick's passion for parachuting turned to obsession and he found himself heading down to Lakewood more and more frequently, Janice and their infant daughter, Donna, would often accompany him. Janice even made five static-line jumps and a few free falls of her own—all under Nick's tutelage, of course. She quietly rooted him on during skydiving competitions and claims that twice during that period he saved fellow jumpers' lives during free-fall mishaps.

"He would just take me over to the airport," she recalls, "and he would explain it all to me. Every inch of the way, he would explain what he was doing. And he would plead for my cooperation. It was important for him that I be okay with it. He showed me every aspect of it, including how safe it was."

Eventually Nick and Janice agreed to leave Union City for good and move down somewhere around Lakewood so that Nick could be nearer the drop zone. Janice liked the idea of moving to the shore and thought it would be a healthier environment in which to raise a family. As it happened, the Lakewood jumpmaster knew of a place for rent in nearby Brick Township, and in the winter of 1963 Nick and Janice took possession of a small house on Nottingham Drive with a little yard on a quiet oak-lined street just a couple of miles from the sea. More important, it was a mere ten-minute drive to the airport.

Whenever he wasn't at the drop zone, Nick lost himself in research and planning. No one remembers him ever doing anything so mundane as mowing the lawn or fixing things around the house, nor did anyone ever catch him lazing around on the sofa. His father would occasionally stop in to take care of maintenance matters and various chores. Janice appreciated Florenz's handyman visits, but it's possible that Nick himself never even noticed.

In addition to the big project, Nick continued to work on his TV documentary idea, the planning for which would serve as a bridge from the wild world of jungles and dangerous animals to the even wilder one of high-altitude skydiving and aeronautics. The project's mission statement testifies to Nick's increasing sophistication in packaging and selling his ideas, a skill that would serve him well as his ambitions grew: "To produce, with extensive use of a single-engine light plane, a series of 13 or 26 half-hour adventure TV programs, revealing various phases of adventure through aviation. A modern approach to modern adventure, in an attempt to unfold an interesting, informative, educational, and entertaining series of programs."

His notes included plans to pitch the concept to each of the major networks and funding strategies that included approaching government officials in tourism-minded countries where the filming would take place. By the time of the move to Brick Town, Nick had written treatments for the first twelve episodes, beginning with what he knew best. Significantly, the seeds for what would later become Project Strato-Jump were clearly germinating by this point.

1. Venezuela (Angel Falls, Devil Mountain)
2. Brazil (Amazon Valley; segment to include "piranha and boa constrictor feeding")
3. U.S. Wildlife (saltwater crocodiles in Florida; alligator and gar hunting in the Louisiana swamps; rattlesnake hunting in Oklahoma; Gila monster hunting in Arizona)

4. Soaring (including attempts on the sailplane altitude and distance records)
5. Ballooning (including attempts on the hot-air altitude record and on the U.S. sport parachuting record following a jump from a gas balloon)
6. Skydiving (including various record attempts and a stunt: a jumper on water skis towed into the air by a sea plane, followed by a 10-second free fall)
7. Mexico (including alligator and poisonous snake hunting)
8. Central America
9. Peru/Colombia (Indians, shrunken heads)
10. Galapagos Islands
11. Andes Mountains
12. Single-engine airplane altitude record attempt

As Nick's focus on parachuting intensified and his jump logbook filled up, he learned of a U.S. Air Force officer named Joseph Kittinger who had bailed out of a gas balloon at a mind-boggling 102,800 feet back in 1960. Kittinger's descent had not been a pure free fall—his rig had included a small drogue chute designed to stabilize the jumper and prevent dangerous spins—but Project Excelsior had been focused on emergency escape procedures and equipment rather than on world records, so the Air Force had never sought official validation for Kittinger's jump. Nick had also begun to learn more about the Soviet jumper who held the official parachute altitude record. In November of 1962, Maj. Eugene Andreev had jumped from a balloon called the *Volga* at an altitude of 83,524 feet, dropping 80,360 feet without the use of a drogue chute or any other kind of device to stabilize himself in free fall. While Andreev had not gone as high or fallen as far as had Joe Kittinger, his pure free fall, which had been duly witnessed by the Fédération Aéronautique Internationale (FAI), had given him the official international record.

Nick immediately understood that one of the biggest problems in putting together a world-record jump would be raising the funds necessary to acquire the hardware, crew, testing facilities, and launch services for a super-high-altitude balloon flight. He had no way to estimate reliably what it might cost, but he suspected it would be expensive. To augment his business income, he got himself and Vern, home on vacation from his college teaching job, hired as iron workers on the Verrazano Narrows Bridge construction project.

Not much later, he found a buyer and sold Animal Wholesalers, Inc., and took a job driving big trucks at night (often long-haul trips to Canada) so that he could devote his daylight hours to research and planning.

"Sometimes he would take long runs," Janice says, "driving the trucks three days and nights in a row, then he'd take three days off. And those days he would spend working. Always working. He didn't need a lot of sleep."

Nick was a quick study. If he came across something he didn't understand or a problem he couldn't immediately solve, he'd analyze it, break it down into components, and grind away until he had gained sufficient leverage to neutralize it. He collected volumes of technical reports, magazine and newspaper articles, brochures and manuals, and organized it and filed it all away for future reference. While he accepted the fact that he would not have time to master each of the myriad facets of his project, he learned to assimilate difficult concepts much as a corporate CEO understands the fundamentals of his business, if not the full technical detail.

From this point, it would all be pure improvisation. He would have to invent his universe out of whole cloth. Yet for Nick Piantanida, it was a time of untethered hope, and nothing within his vision, not even the sky itself, was the limit.

2 SPACE, INC.

Whether Nick was aware of it or not, aeronautics had a rich tradition in New Jersey dating back more than 130 years. In fact, Charles Ferson Durant of Hudson County had been the first great American balloonist. Durant's landmark flight on September 9, 1830, had taken him from his launch spot at Castle Garden on the southern tip of Manhattan Island to South Amboy, New Jersey. Durant shared more than just geography with Nick: he was constantly embroiled in offbeat and occasionally ill-considered schemes, he was fascinated by exotic animals, and he was as obsessed with the deep oceans as he was with the extreme skies. He also had a knack for self-promotion. Moments before lifting off on his "Grand Aerostatic Ascension," Durant treated the 20,000 New Yorkers who had shown up to see him off to a dramatic reading of his own poem, titled "The First American Aeronaut's Address." It began: "Good bye to you—people of Earth / I am soaring to regions above you." But the most important thing Durant and Piantanida would share was their mode of transport. Nick had come to understand that the gas-filled plastic balloon was the vehicle he would need to reach the altitudes Andreev and Kittinger had visited. No civilian aircraft was capable of those heights, and Nick wasn't going to be able to scam a high-performance jet from the Air Force no matter how far he cranked up the charm. Besides, at the speeds stratospheric planes traveled, just getting out in one piece qualified as something of a miracle. Balloons were the perfect platform for a record-setting parachute jump and, compared with the alternative, they were both safe and affordable. While the preferred lifting gas would now be helium rather than the flammable hydrogen used in the nineteenth century, the basic concept and mechanics of Nick's balloon would be not much different from Durant's.

The earliest written plans for Nick's project envisioned three stages, and an extremely aggressive schedule:

1. A jump from 20,000 feet by June of 1964.
2. A jump from 43,500 feet by August of 1964.
3. A jump above 80,000 feet sometime after September 1964.

The plans called for all three flights to launch from and terminate on New Jersey soil. Interestingly, at this point Nick was still considering the use of a partial-pressure suit like the Air Force balloonists of the 1950s had used, the same style Kittinger and Andreev had worn on their record jumps.

But as his ambitions expanded and he began to consider a jump from somewhere well above 100,000 feet, he decided that he would be wise to wear a full-pressure space suit similar to those worn by the Mercury and Gemini astronauts. The bulkier suit would surely complicate the business of skydiving, but should in theory provide a greater measure of protection against the unforgiving elements.

Nick collected what little reliable information he could locate about the behavior of falling bodies in the near-vacuum of the upper atmosphere and contemplated some of the difficulties those before him had encountered: accelerating spins, carbon-dioxide poisoning, overheating, blinding sun, stratospheric storms. But there's no indication that anything he came across cast much of a pall on the grand dream of a world-record free fall. All the obstacles appeared to be surmountable given the right hardware, the right people, and—of course—the money.

Nick wrote in his log: "I reasoned that except for equipment, there really shouldn't be any great problem. Mathematically it worked out that the amount of pressure being forced against a falling body should be equal at any altitude once terminal velocity has been reached. Speed is increased tremendously at higher altitudes, but this is due to the decrease in atmospheric pressure. The motivating force (1G) remains constant, thus accounting for the equal amount of pressure against the body."

Just as he'd taught himself the rudiments of rock climbing from library books as he trained for Angel Falls, it was to aviation history and physics textbooks he now turned. From this point forward, Nick would spend as much time reading and thinking as he would hurtling through the air. He was entering a new phase of his life in which he would finally apply himself to a course of study with an enthusiasm he'd been unable to generate in his half-hearted collegiate experiences. Analyses of the astronaut selection process at NASA had lately suggested that space programs would do far better to focus their recruitment efforts on the graduate schools than on the varsity backfields or boot camps: scientists rather than athletes or warriors. And Nick—with the Ichi-Ban, the basketball, and the backyard cobra fights now behind him—was setting out, with a great and largely unfamiliar discipline, to transform himself into the director of a one-man aeronautical research program.

Against type, he even bought himself a crisp new business suit and a leather briefcase.

One afternoon in the fall of 1963, Nick managed to reach U.S. Air Force Capt. Joseph Kittinger on the telephone. Fellow test pilots and NASA's Mercury astronauts stood in unabashed awe of Kittinger's 1960 bailout from the edge of space. He and his Project Excelsior team had proved to the skeptics the survivability of emergency escape at extreme altitudes and had done so using the standard-issue suit worn by the Air Force's jet pilots. It was an extraordinary and genuinely heroic achievement. But by constitution, Nick Piantanida was rarely, and only reluctantly, awed by anybody.

"I'm going to break your record," he told Kittinger matter-of-factly. "I'm going to fly higher and fall farther."

Kittinger, however, was suspicious and immediately lumped Nick in with the other daredevils and glory-mongers who'd contacted him since his historic jump. "Wannabes," Kittinger called them. He grilled Nick for a while, posing some hard but germane technical questions about pressure-suit performance and altitude training, probing at the extent of Nick's practical and theoretical knowledge, and challenging a number of his assumptions. Kittinger took pains to impress upon Nick just how difficult and dangerous such a flight would be. On a project like this, he explained, pilot survival didn't just happen. There were forces in the stratosphere that could kill you instantly. If you didn't understand that or weren't properly prepared for it, you had no business up there.

Nick shrugged off the warnings. "I'm not too worried about any of that," he said, which brought him to his motive for calling. "I'd like to ask for your help," he told Kittinger. "You're the expert. I want you on my team, to be part of my project. Will you join me?"

For Kittinger, this was the easiest decision he'd ever made. The answer was a firm no. Apparently immune to the powers of the Piantanida persuasion, he wished Nick luck and forgot about him, assuming—as with the other wannabes—that it would be the last he would ever hear from or about this headstrong civilian.

Altitude

By the time the winter of 1963/64 had begun to loosen its grip on the northeastern seaboard, Nick—his determination solidly intact—was back at Lakewood racking up jumps. He had by this time confided his dream to a handful of friends, but Janice was still the only person who knew the true extent of Nick's building obsession. She could see that this thing was growing into something big, and she knew Nick well enough to understand that he could never let it drop now. It had already gone much too far, and it spooked her to think where it all might end up.

What he really needed at this point was some hands-on high-altitude experience. But opportunities to parachute from above 20,000 feet were rare—especially for a newcomer to the sport. In April Nick learned of a 30,000-foot jump being promoted by a Lakewood instructor named Bob Spatola. He tried to talk his way onto the jump team, but his relative inexperience, he was told, made it impractical. Nick persisted. Finally, Spatola relented and agreed to consider Nick for the jump under one condition: he would have to earn his Class C rating prior to the scheduled May 23 jump date. Nick was elated, until he checked the Class C requirements and found he'd need seventy-five total free falls to earn his certificate. He only had twenty-five to his credit.

Over the next few weeks he jumped like a maniac, making ten jumps in a single day on one occasion. He started skipping work so he'd be fresh each morning and ready for the first jump load of the day. When the sun rose on May 22, the day before the big jump, Nick was still four free falls short of his goal. He rented a plane at the Caldwell Airport and flew to Long Island to retrieve some high-altitude gear he'd arranged to borrow for the 30,000-foot jump. He doubled back to New Jersey for two jumps before heading off to yet another airport to pick up an oxygen mask. When he found a partial load of skydivers going up at Cameron Air Park, he hopped onboard and made another quick jump. Then he hurried back to Lakewood in time to catch the last load of the day there and made his seventy-fifth free fall—qualifying for Class C.

But there was a problem. He still had to ferry the rented plane back to the Caldwell Airport, and by this time it was dark. Nick had never soloed at night before, and it had been nearly a year since his last dual night flight. He climbed in and took off from Lakewood with a knot in his stomach. He flew a heading toward Caldwell, but had radio trouble and was forced to navigate visually. Luckily, he could just make out the Empire State Building in the distance and, with that as a reference, was able to locate Route 46 and get his bearings. He recalled an incident some years earlier in which a pilot in distress had made an emergency landing on the George Washington Bridge, and he noted the location of the bridge in case he ran out of better options. Nick made the airport and landed safely, but because in his frantic rush he'd neglected to refuel, the engine quit before he could taxi all the way in and he was forced to push the plane by himself the final yards to the flight line.

Unfazed by the close call, Nick was up the next morning before dawn and on his way—more than three hours by car—to Millville in southern New Jersey down near the Delaware Bay where he was to meet Spatola's team for the jump. He had never jumped above 15,000 feet, and he was jacked up. When he got to the airport at Millville at 7:30, the skies were blue. But by the time all the preparations had been completed, the gear loaded, and the B-26 that was set to haul the parachutists up to altitude was fueled and checked out, it was 9:00 and big clouds had started to slide in. By 9:30 the clouds had darkened and the wind had kicked up to 25 miles per hour. Minutes later, the jump master canceled the flight.

It was only the first of many disappointments to come in the busy months ahead.

A couple of weeks before the canceled Millville jump, Nick had written to the Air Force requesting information on high-altitude balloon and parachute activities. He had heard stories and seen a couple of magazine pieces about military research programs from the 1950s that had done amazing things in the stratosphere. When he returned home from Millville, he found a letter waiting from Lt. Col. Bernard Peters of the Air Force Systems Command, Washington, D.C. Peters confirmed that the Air Force had in fact been involved in manned and unmanned plastic balloon flights for some seventeen years and held a number of unofficial altitude and parachute records. Peters's letter contained a fair amount of misinformation with regard to specific records and dates, but for Nick it was a start.

More helpfully, Peters suggested that Nick contact the FAI in care of the National Aeronautic Association (NAA) for a report on current official records. (The FAI delegates its authority to the NAA for record attempts made in the United States.) He also suggested that Nick write to Otto Winzen, the flamboyant president of Winzen Research, Inc., in Minneapolis. Winzen's company, which he'd run with the help of his capable young wife, Vera, had built massive polyethylene balloons for the Air Force's and Navy's manned high-altitude projects. Winzen had also authored a celebrated paper in defense of balloon-borne research in the space age that had been delivered at the eighth International Astronautical Congress in Barcelona in 1957, and Peters advised Nick to familiarize himself with it.

A week later, another letter from the Air Force appeared in Nick's Brick Town mailbox. This one was from Floyd Clark, a second lieutenant at the Missile Development Center at Holloman Air Force Base in New Mexico where several of the Air Force's scientific balloon projects had originated. Clark recommended a book titled *Man High* by Lt. Col. David Simons. Dr. Simons had served as the project officer for the Manhigh program and had been the pilot on the project's record-breaking medical-research flight in 1957.

Don Piccard (*right*) in 1962: Nick's link to the glory days of American aeronautics. (Courtesy Jim Winker)

Nick immediately sat down and wrote to the NAA and to Otto Winzen. He kept copious notes and was determined to follow up every lead. He devised a file system to organize and keep track of what he hoped would become an exhaustive collection on stratospheric flight, balloons, parachutes, oxygen equipment and survival systems, high-altitude physiology, and anything else that seemed even remotely relevant to the dream. He commandeered the spare bedroom on Nottingham Drive and began stacking up boxes full of papers, piles of parachute hardware, and scientific reference books.

Piccard, the Senator, and the Cooler King

At about this time, the summer of 1964, Nick made an important contact in the American ballooning fraternity. His reading had revealed that the first successful manned balloon flights into the stratosphere had been made in Europe by Swiss physicist Auguste Piccard and his colleague Charles Kipfer. Auguste's twin brother, Jean, had emigrated to America where he, along with his wife Jeanette, had made their own stratospheric flight to 57,579 feet in 1934. When Nick learned from a pair of magazine articles—one in *National*

Aeronautics and the other in *Popular Mechanics*—that Don Piccard, the youngest son of Jean and Jeanette, was also a prominent balloonist, he quickly tracked him down. The magazine pieces had chronicled test flights in the earliest hot-air sport balloons, flights that had been arranged and supervised by Don.

Piccard, as intellectually curious and adventurous as his father and uncle, had been a cofounder of the Balloon Club of America, a small group of Pennsylvania-based aeronauts who were active in the decade following World War II. Two years before being contacted by Nick, Piccard had established an altitude record for a 250-400 cubic-meter balloon and was now in charge of sales and promotion for the fledgling sport balloon program at Raven Industries, a design and manufacturing facility in Sioux Falls, South Dakota. Raven's S-series models were designed specifically for civilian sport flying.

Piccard was intrigued and receptive to Nick's initial inquiries, and agreed to consider building and launching a gas balloon for the record free-fall attempt, although they had yet to discuss specifications or money. But the Piccard contact was crucial for an entirely different reason: it gave Nick a prestigious and credible name to drop when schmoozing other members of the ballooning community. He could immediately begin announcing that Piccard was on board, an opportunity that he would rarely miss taking advantage of in his talks with reporters and prospective donors in the months to come.

Through it all, Nick needed to continue his parachute training. There was still much to learn. In early June, he made a 20,000-foot jump from a Cessna 205. A week later, he jumped from 23,500 feet. He tried to talk his way onto every high-altitude jump load he could find, but between bad weather and mechanical problems, precious few opportunities materialized. Jumps from these heights are a brutal business, even for the properly prepared and well-outfitted. The most immediate concern is the extreme cold. At free-fall velocities in the upper atmosphere, poorly outfitted jumpers can quite literally freeze to death before reaching the ground. That the punishment endured on these high-altitude jumps failed in any way to dampen Nick's enthusiasm—as they would do to the ambitions of many who would attempt to follow in his footsteps—is testament to his dedication.

On June 15 Nick wrote to the Aerospace Medical Division at Brooks Air Force Base in Texas requesting information on Project Manhigh, and asserting—rather cavalierly—that he intended to surpass both Andreev and Kittinger and set a new world free-fall record. The letter was eventually forwarded to Dr. David Simons who would turn out to be Nick's first promising contact with a member of the prestigious Air Force balloon fraternity. Simons graciously forwarded copies of the technical reports on two of the Manhigh flights and wished Nick luck.

In the course of his reply, in which Nick predicted—again, rather unrealistically—that his attempt on the world balloon altitude and free-fall records might occur as early as the fall of that same year, he did his best to convince Simons that his was a well-thought-out, serious project, writing, "this isn't just another gas-bag dream." Nick went on to request further technical detail on the clothing and equipment Joe Kittinger had used on his record free fall, unaware that Kittinger's jump had been made under the auspices not of Simons's Manhigh but of Kittinger's own Project Excelsior. There's no way Nick could have known of the personality clash that had led to a rift between Simons and Kittinger during Manhigh. Unfortunately, that was the end of any future cooperation he might have received from David Simons. Like other veterans of the scientific balloon programs of the 1950s, Simons was always hesitant to get involved with civilians. At some level, he distrusted all would-be record-setters on the face of it—Kittinger included.

"I understood that impulse, to set a record," Dr. Simons would say. "But that was never my motivation. I was interested in scientific discovery."

The Lakewood jumpers were a generally accomplished lot, but that summer Nick made the acquaintance of perhaps the most remarkable of them all. Jackson Barrett (Barry) Mahon was an enthusiastic parachutist who also made movies. His, however, was hardly the typical Hollywood résumé. Born and raised in Southern California, Mahon learned to fly airplanes while in high school in the late 1930s. Underage and unwilling to wait for his country to join the war in Europe, he signed on with Britain's Royal Air Force in 1941 and flew with the American-volunteer Eagle Squadron. Mahon completed ninety-eight missions and shot down or damaged several German fighters, becoming the squadron's fourth ace during the Operation Jubilee commando raid on Dieppe in August of 1942 during which his own plane was shot down. The British Air Ministry awarded him the Distinguished Flying Cross for his heroism.

While a prisoner of war, Mahon attempted several escapes and helped architect the escape tunnels at the Stalag Luft III camp in Polish Silesia where he became known as "the Cooler King" for his frequent stints in the camp's solitary confinement cells. Steve McQueen's character in the 1963 movie *The Great Escape*, adapted from a book by Australian flyer Paul Brickall, was based largely on the bravura prison-camp exploits of Barry Mahon.

After the war Mahon became the personal pilot—and eventually business manager—for actor Errol Flynn. He was a producer on films starring Flynn and Gina Lollobrigida and later directed a string of his own B-movies with titles like *Rocket Attack, U.S.A.* (1958), *Violent Women* (1959), *Cuban Rebel Girls* (1959), and *Prostitutes Protective Society* (1966). These were not, obviously, Academy Award material. An ungenerous reviewer assessed one of Mahon's productions this way: "Bad acting, bad sets, bad plot, bad premise, bad sound, and bad cinematography add up to a profoundly bad film." Mahon and his company, Mahon Productions, were also busy during the late fifties and early sixties profitably cranking out films that were more marginal yet: hard-core pornography loops for the adult-only market in New York City.

Nick had met Mahon during one of his early trips to the Lakewood Center and had convinced him that a feature-length film of the world's longest free fall had genuine commercial possibilities. Mahon had the expertise, the equipment, and the crew, as well as the unique background in both aviation and filmmaking. Mahon himself was a delightfully worldly individual with a great sense of humor and a genuine swashbuckling *joie de vivre* that naturally had great appeal for Nick.

Mahon immediately took a keen personal interest in Nick's project and offered to help—at his own expense—in any way he could and suggested that in addition to any cinematic contributions he might also be able to assist with fundraising. He even volunteered to exploit some contacts he had at B. F. Goodrich—a pioneer in early pressure-suit design—to try to get the company to supply a space suit for Nick's free fall.

Meanwhile, Nick continued jumping and working to sharpen his skills as a skydiver. On July 4 Lakewood director Lee Guilfoyle, who had been one of Jacques Istel's partners in the founding of Parachutes, Inc., hired Nick as an instructor. It was a highly desirable position with plenty of perks and one that offered real status in the local community.

"The average instructor at Parachutes, Incorporated had great prestige," Jacques Istel explains. "It was something like being a ski instructor in the early days. The biographies of all the instructors included a fair amount of adventure of one type or another. And if you do unusual things like this, you become comfortable enough in your own skin not to be concerned about what people think. Even most of those who were awed by the instructors probably thought they were nuts."

Nick had 110 jumps to his credit by this point, a tally that he would more than double by the end of the year. But he still needed additional high-altitude work. More specifically, he needed high-altitude experience wearing a pressure suit like the one he intended to use somewhere above 100,000 feet. The aerodynamics involved in free fall with a bulky, restrictive full-body enclosure were no doubt going to require changes in technique, changes that could only be worked out in actual jump trials.

In the mid-1960s the U.S. Army Parachute Team was still testing partial-pressure suits—which use inflatable bladders to "pressurize" only selected parts of the body—in decompression chambers to simulate high-altitude jumps from 75,000 feet. But in spite of Air Force programs that had years earlier proved the usefulness of those suits to even higher altitudes, Nick was convinced that only a full-pressure garment like those worn by astronauts would provide the protection he would need for an unstabilized jump from the upper stratosphere. Acquiring a full-pressure suit, however, was turning out to be even more difficult than getting access to high-altitude aircraft. Only a handful of manufacturers were capable of constructing these suits, and they were all military contractors unaccustomed to and uncomfortable dealing with civilians. Nick made his pitch to baffled or indifferent audiences at firms like Garrett Airesearch in Los Angeles and Airesearch Manufacturing Company in Phoenix. Barry Mahon's contacts at B. F. Goodrich had been of little help. Goodrich's product manager had written Mahon explaining why the company was refusing to sell Piantanida a pressure suit: "Without the backup of the years of experience and technical capabilities such as are available in the military services, we cannot be confident that a venture of this nature will be successful. Even with the military's extensive experience and the unlimited funds and availability of technical personnel, periodically, programs meet with failure. I am sorry that we cannot cooperate in this venture."

Nick would receive a different, but equally discouraging, reply from Arrowhead Products in California. The project manager of Arrowhead's Life Support Systems Division explained that the problems of extreme high-altitude bailout are daunting even for specialists, and that solutions are expensive, perhaps prohibitively so, for an individual. He strongly suggested that Nick contact the U.S. Navy for whom Arrowhead had been a contractor. Like the Air Force, the Navy had sponsored a number of high-altitude balloon flights and had helped NASA test pressure suits designed for American astronauts. "I am not attempting to discourage you," the company's representative wrote, accomplishing precisely that, "and we will quote a suit if you still desire. The price would be considerable."

This was not good news given Nick's almost total lack of financial resources at this point. It was obvious that he was going to need to cultivate sources for major funding if he was going to be able to move the project forward, and it seemed likely that he was going to have to develop some better contacts within the military community who could help him get his hands on a pressure suit.

What Nick needed was a friend in high places and, by the end of July, he had managed to gain the ear of his local township committeeman, Albert Cucci. It was a start. Cucci consequently got in touch with Harrison A. Williams Jr., U.S. senator from New Jersey, and asked for help convincing the Navy to cooperate with the project. At Nick's suggestion, Cucci framed his request with the news that a successful jump would capture a major aeronautical record then held by the Soviets.

Williams was immediately enthusiastic. Within a week he had buttonholed his own Navy contacts on Nick's behalf and assured Cucci that the Navy would cooperate: "I might mention that the Navy has informed me that they usually try to be as helpful as possible concerning matters of this type—especially since we are competing against the Russian record." Nick's strategy had worked. It was the beginning of an association that would ultimately pay huge dividends. Over the following eighteen months, Williams and his staff would spend many hours helping Nick navigate the unmapped back roads of American military and industrial bureaucracies in the attempt to acquire the survival equipment necessary to operate in the space-equivalent conditions twenty miles above the earth.

Williams made inquiries with the Air Force to see whether Joe Kittinger might be persuaded to change his mind and offer his services to Nick's project. Nick continued to believe that if only Kittinger could be convinced that the project was truly a serious one and that Nick himself was both capable and willing to do whatever it would take to prepare for his mission, that the Air Force's most celebrated aeronaut might reconsider. There just weren't that many people who knew what it took to run a high-altitude manned balloon program, and the few contacts Nick had made in the ballooning world were adamant that Kittinger's experience and knowledge would be the surest hedge against failure.

Later that summer, the Air Force Surgeon General's office phoned Kittinger, asking if he would be willing to reconsider and volunteer for the assignment. Kittinger, annoyed that Nick had refused to take no for an answer the first time and convinced that it could be nothing but a no-win situation for the Air Force, refused to budge. If he got involved and Nick was successful in breaking any records, he reasoned, the Air Force would get none of the credit. Yet if Nick succeeded only in killing himself, the Air Force would surely get all the blame. For a second time, Kittinger made clear his intention to have nothing whatsoever to do with Nick Piantanida.

In spite of the double snub from Kittinger, total silence from Otto Winzen, and a polite brushoff from David Simons, Nick was managing to make some small progress. In addition, he still had the attention of Don Piccard. In September Piccard applied some creative thinking to the pressure-suit problem and proposed a novel idea: a pressure vessel. In the latter days of World War II the Army had experimented with "pressure bags" into which B-29 pilots could zip themselves in the event of cabin depressurization. It was described as a "get-the-pilot-down-quick" system. Piccard's concept was essentially a hardened, aerodynamic pressure bag with parachute attached. He described it in a brief letter to Nick: "A four foot sphere with fins could carry you up, be released and then fall in a stable condition until parachute deployment. It would be a lot cheaper than a suit and a lot simpler. Perhaps even a lot safer."

But a whole lot less interesting to an accomplished parachutist. Piccard's idea was a nonstarter. The dream was about free-falling through the stratosphere, employing the skydiving techniques Nick had spent the past year perfecting. It was about following in the footsteps of Andreev and Kittinger. The notion of dropping inside some sort of metal container must have sounded like plunging over Niagara Falls in a barrel. Regardless of how free fall inside a pressure vessel would be regarded by officials at the FAI and NAA, and whether or not such a drop would even qualify as a legitimate world record, it's easy to understand Nick's immediate and outright dismissal of any such notion.

The Hustler and the Con Man

Next to finding a supplier for the pressure suit, funding remained the most daunting issue standing between Nick and the fulfillment of his dream. A New York fashion photographer who had once used Nick as a model was a constant source of ideas on how money could be raised; more important, at least at the moment, the photographer contributed $1,200 worth of seed money. Roughly half of that amount went to the purchase of parachuting gear and to cover the expense of continued jump training. The rest was used to develop a direct-mail fundraising campaign. Nick purchased a "select" list of 10,000 names from a public relations firm and composed a solicitation letter. He was now using the name "Project Strato-Jump" and hoped that the letter's appeal would benefit from this military-sounding designation. Over the course of two nights Nick and Janice worked through the wee hours stuffing, addressing, and stamping envelopes that were mailed to recipients in every state of the Union. The results? One dollar sent in by an elderly woman in Florida.

A second mail campaign fared better. It was constructed as a patriotic appeal to the American people and mentioned the Russian record more prominently. It was also, like most of Nick's marketing efforts, a bit optimistic as concerned the status of the project considering that Strato-Jump still had neither balloon nor pressure suit: "All the operational phases have been completed and, with the exception of financial support, everything is ready to go. Various government agencies are very interested in Project Strato-Jump, but unfortunately money cannot be appropriated for its use.

Because of this, we are appealing to you, the American public, to become a part of this important project." The letter concluded with a description of a new sponsorship award program. Contributors of $20 got a certificate "suitable for framing." Contributors of $25, $50, and $100 received "appropriately sized" wall plaques. And any individual or organization donating $100 or more was given a "handsome desk trophy" that consisted of a cube of transparent Lucite with the arched figure of a skydiver suspended within. For months, boxes of trophies lined the walls of the house in Brick Town. "Please make checks payable to Project Strato-Jump," the letter concluded. It was signed by "E. I. Patterson, Project Secretary." But of course Nick couldn't actually afford a secretary. E. I. Patterson was, in reality, his mother-in-law, Ida.

Nick's own family was never so supportive. In spite of his efforts, he had never been able to offer an explanation for his motivations that they could understand. He tended to counter the doubts with a question of his own: "Why did Lindbergh fly the ocean?" But that did little to assuage the home-front. They knew too much to believe the line Nick would use regularly with newspaper reporters, that his role in Project Strato-Jump would be "safer than rolling out of bed."

One evening when Nick and Janice had driven up to Union City to have dinner with Nick's parents, Catherine lured Janice down to the basement. The older woman was in tears, pleading. "Please, you have to tell him not to do this thing. This is terribly dangerous and I don't want him to do it."

Florenz not only refused to discuss the matter of Project Strato-Jump, he would often leave the room if the subject so much as came up. Vern seems to have mostly just avoided the issue, hoping perhaps that in time his brother would move on to a new and more sensible scheme.

No one can accuse the family of overreaction. Nick was a safety-conscious jumper, one of the first at Lakewood to purchase and use an automatic reserve chute opener. But as he continued to pad his jump total that fall, an incident occurred that underscored the very real risks inherent in jumping from even relatively low altitudes, let alone from the edge of space. Don Sullivan was a Lakewood regular and one of the more passionate skydivers anyone had met. Wiry and always full of energy, Sullivan was famous for rushing to squeeze in that one last jump before the sun went down. His parachute rig included an old Army-surplus B-4 harness, a cheap, poorly designed piece of gear that had been generally superseded by Pioneer Parachutes' sport harness a couple of years earlier. Nick and Don had jumped together many times throughout 1964, and they made two jumps on Sunday, September 6, both from 3,700 feet, during which they practiced relative work, briefly hooking up in formation during free fall. A note in Nick's parachute logbook read: "Fun jump—hooked up out of plane."

The next morning Nick and Don jumped again. A few days earlier Don had reported a "hard pull," suggesting some problem with his chute's deployment system. Several people at Lakewood had inspected his rig without finding the cause of the difficulty. But on this Monday morning, 2,000 feet above the ground, Sullivan found himself completely unable to pull his ripcord. Phil Chiocchio, just sixteen years old and the youngest parachutist at the drop zone, had befriended Sullivan. He watched helplessly from the ground.

"I saw him spinning," Chiocchio recalls, "both hands pulling the handle out in front of his face, but no chute. Finally he pulled his reserve chute around 1,000 feet or so." Sullivan's reserve streamered, meaning it released but didn't inflate. Somehow a hitch had formed in the shroud lines. For about ten seconds Sullivan struggled frantically, pulling at the lines in an attempt to open the canopy. "He went in near the high-tension power line next to the New Jersey Parkway. When I got over there an ambulance was arriving and he was lying there conscious and joking. He said to sign him up for a flight later in the day."

Don Sullivan died at a local hospital of internal bleeding from a punctured lung. Some of the Lakewood staff believed he might have survived had he not been forced to wait nearly two hours in the emergency room before receiving proper treatment.

Don Sullivan's death due to equipment malfunction was neither a bizarre nor an unheard-of type of accident. Nor did it inhibit normal activity at Lakewood. In fact, in spite of the fatal incident, Nick—who had gotten fairly close to Sullivan—went ahead and jumped three more times that same Monday. His logbook contains the following rather bland annotation: "Sullivan killed this morning. Didn't feel much like jumping today. Just fell flat and stable on all three jumps." In these early days of sport parachuting, you were never more than a kink or a twist away from the end. Everyone at Lakewood had seen friends die. And they were all very aware that this was one of the prices you paid for a life of free fall.

Piccard was now telling Nick that he could probably arrange to have a suitable balloon built and launched for a fee of $10,000. While this was merely Piccard's estimate and not a formal quote on behalf of Raven Industries, it seemed like a very reasonable number. At the same time, Nick was becoming increasingly frustrated in his efforts to generate cash. He began to sketch out plans for homemade oxygen and communications systems, doing his best to come up with low-cost contingency plans in case it turned out to be impossible to line up the necessary hardware and ensure cooperation from the firms who had the expertise.

But he kept returning to the real sticking point: a pressure suit. He couldn't do that one himself. He was going to have to find some real money and a source for a suit or the dream was effectively dead.

Nick's most valuable connection from a business standpoint would turn out to be with none other than Jacques Istel himself, the man responsible for the birth of skydiving in the United States and for the Lakewood Center. In addition to introducing free-fall techniques to America, Istel had designed and developed the nation's first sport parachute rig, the original model of which now belongs to the Smithsonian Institution.

Istel was a distinctly odd duck. With his aristocratic French accent, barrel chest, and jutting jaw, he was an unforgettable—and frequently controversial—character. But Istel, as aggressively confident and daring as they come, saw a kindred soul in Nick Piantanida and agreed, if reluctantly at first, to help him with the financial aspects of his dream.

So, as Nick discovered the difficulties of getting other people to pledge their money to an unknown civilian who wanted to break one of the most difficult and exotic records in all of aeronautics, Istel helped him form a suitable business entity—SPACE, Inc.—headquartered in the Time/Life Building in New York City at the office of another Istel concern, Intramanagement, Inc. Istel and Piantanida named themselves president and vice president, respectively. Istel contributed $10,000 of his own funds—or, rather, the funds of Parachutes, Incorporated—to SPACE (Survival Programs Above a Common Environment). As one who knew Nick well in the mid-sixties observed: "Nick had what it took to get money out of people. He was a hustler. Nobody else could have done it. It was his drive, his ambition, his determination, and his ability to make it happen."

The same could justifiably have been said of Jacques Istel. "Basically," Istel would confide years later, defiant and with a proud gleam in his eye, "I am a con man."

When, one morning in late 1963, Nick had first burst into the offices of Intramanagement, Inc. with an introduction from Barry Mahon, Istel's first reaction to Nick's sales pitch had not been terribly encouraging. "Terrific!" he had said, the notion of a recordbreaking adventure in the stratosphere appealing to his innate love of adventure. "Just terrific!" Then the practical side of Istel's nature emerged. "But not for us [Intramanagement, Inc.]." Istel could see immediately that the project would be costly and difficult, and also perhaps that the chances of success were something less than an even-money proposition. "If you're going to do it," he continued soberly, "you'll have to find some asshole to finance it."

And that was how the first meeting ended, with the con man turning the hustler down flat. But less than a year later, here Istel was signing a five-figure check that would—they both hoped—go toward the purchase of a giant balloon that could be used only once. If Nick could persuade a hard-nosed businessman like Istel to part with that kind of cash for something as ephemeral as a plastic balloon, he was surely on his way to the stratosphere. As he would confide to a reporter, "It's all promotion."

It was the triumph of the hustler.

From the moment he first learned of Piantanida's high-altitude parachute project, Democratic Senator Harrison Williams—one of big labor's best friends in Congress—embraced the idea of a New Jersey man, and not just any man but a truck driver, a Teamster, conquering the stratosphere and displacing a Russian in the record books. From mid-1964 through early 1965 it seems as if Williams's office worked almost full-time for Nick, in spite of the fact that the senator was in the middle of a reelection campaign for much of that period. In fact, Williams went so far as to provide Nick with private office space—not much more than a closet, but with a desk and a telephone—in the Senate Office Building on Capitol Hill so that Nick could make the most of his time on his trips to Washington. This also allowed Nick to approach potential funding and equipment sources with the line, "Nick Piantanida calling from Senator Williams's office."

The big issue at hand was a full-pressure suit and how one could be obtained for civilian use. But even for an influential member of the U.S. Senate, the roadblocks were formidable. In August Williams informed Piantanida that the Navy was simply not going to allow him to use one of their suits. Still, Williams was optimistic that Navy facilities and personnel at Norfolk, Virginia, would be made available for training in the event that Nick could manage to locate a suit from another source. Nick got in touch with the Bureau of Naval Weapons at Norfolk who advised him to send a résumé to the Bureau of Medicine and Surgery. When Medicine and Surgery received the résumé, they thanked Nick and informed him that a pressure suit would need to be obtained through the Bureau of Naval Weapons. Nick was getting his personal introduction to the ancient art of the bureaucratic stonewall.

The Bureau of Medicine and Surgery followed up with the news that Nick's request could not be considered until he could supply the results of an electrocardiogram test. Nick promptly got the test and forwarded the "100% normal" results. About a week later, having heard nothing, Nick managed to get an officer from Norfolk on the phone who admitted that his superiors were telling him "hands off" the Piantanida request because "the Navy can in no way benefit by it." It was a carbon copy of Joe Kittinger's no-win theory.

Nick's frustration with the Navy was mounting, but he persevered. On October 17 he wrote to the Parachute Club of America in Monterey, California, in an attempt to enlist that organization in his quest for a pressure suit. He spelled out the particulars of the project in some detail and listed his objectives, announcing that he now intended to launch from South Dakota sometime between January and May 1965. He played up his association with Don Piccard and with Senator Williams. He provided an overview of the oxygen and communications systems he intended to use and declared that he would commission a three-million-cubic-foot plastic balloon. And, of course, he described in some detail the parachute apparatus that had been selected.

In a genuine coup for Strato-Jump, the Pioneer Parachute Company of Manchester, Connecticut, had become the first major outside participant to join the project as a partner. Not only had the Piantanida/Istel team persuaded Pioneer's president, Michael Kagan, to purchase some $12,000 worth of shares in SPACE, Inc. (25 percent of the operation), but the company had offered to provide Nick with a modified high-altitude version of its new state-of-the-art Para-Commander, described as "more of a glider than a parachute." (Pioneer would, over the course of the project, donate approximately $8,000 worth of parachute hardware and engineering services to Strato-Jump.) The PC, still round in shape but with strategically placed drive slots and side stabilization panels, provided a forward speed of thirteen to fifteen miles per hour in calm conditions and a descent rate of only fourteen feet per second for a man of Nick's size. On a leap from the stratosphere, however, given the more than 100 pounds of additional gear Nick would need to carry, the descent rate would probably be closer to eighteen feet per second. But the PC was significantly more maneuverable and forgiving than anything that had preceded it, and it would become the rig of choice for serious skydivers until the advent of modern ram-air chutes.

The last of the six objectives Nick included in his letter to the Parachute Club of America is revealing of his state of mind in the fall of 1964: "To show that the military government, with virtually unlimited funds, doesn't have a monopoly on breaking all the outstanding aeronautical records." For Nick, it was becoming personal. He was continuing to jump every chance he got—he made more than seventy-five jumps at Lakewood from mid-August through mid-October, with an even dozen from 10,000 feet or higher—but the thrill of skydiving wasn't enough to temper the building frustration. The game was no longer merely Nick versus the Russians; it was now Nick versus the entire American military industrial complex.

In late November, having all but given up on the Navy, Nick wrote to Senator Williams, congratulating him on his reelection and suggesting three potential alternate sources for a pressure suit. The first was the civilian Federal Aviation Administration (FAA), which had ample facilities and equipment, and plenty of expertise. The second was the Air Force, which used such suits on a regular basis for some of its high-altitude jet pilots. The third involved combing the military surplus market for one of the Arrowhead suits that had recently been discontinued by the Navy; perhaps, Nick suggested, the Navy would be willing to help him locate such a suit since his use of it would no longer be any responsibility of theirs. Williams made inquiries at the FAA and the Air Force. He also went back to the Navy on the Arrowhead suit idea but got little more than a promise to consider the request; an outright refusal followed a few days later. So it was final. Nick would not be allowed to use any suit that was originally designed on a Navy contract, surplus or not. In addition, Williams was forced to report, his requests had also been turned down by both the Air Force and the FAA.

Meanwhile, Barry Mahon had gone back to the reluctant representative at Arrowhead to find out just how much they would charge for a pressure suit. The answer was a firm $15,000 on delivery, take it or leave it. The price seemed exorbitant to Nick—the total cost of the pressurized suit B. F. Goodrich had built for aviator Wiley Post in 1934 had been a mere $75—and he felt he had no choice but to keep looking. He called contacts at the University of Illinois to inquire about the use of their decompression chamber for physiological training but was unable to get a straight answer. He asked officials at Edwards Air Force Base in California for the loan of two barographs for recording altitude but was shuffled off to the NAA. He borrowed fifty dollars from Mahon and drove to Washington, D.C., to meet with officials from the NAA who were upbeat and encouraging about the prospective record attempt but less sanguine about Nick's odds of getting his hands on a suitable pressure garment. There was just no precedent, they explained, for these sophisticated survival systems being used for anything other than official government business.

Nick was neither the first nor the last civilian airman to have his plans frustrated by the stubborn reluctance of the American military and its contractors to make a pressure suit available. In the early 1930s a New England adventurer named Mark Ridge announced his intention to attempt a stratospheric flight in an open gondola balloon. Ridge contacted the Army Air Corps, the Navy, and Harvard University seeking assistance in the design and manufacture of a pressurized system capable of protecting a high-altitude aeronaut. When all of his requests were denied, Ridge approached Dr. John Scott Haldane in London; Haldane had developed a revolutionary pressure suit for deep-sea divers some thirty years earlier. Haldane and colleague Sir Robert Davis of Siebe, Gorman, and Company modified a pressurized diving suit and Ridge wore it in Haldane's decompression chamber to an altitude equivalent of 90,000 feet. This was the first simulated trip into the upper stratosphere and the suit performed flawlessly. To Ridge's great dismay, however, the British Air Ministry denied him permission to wear the suit in an actual stratospheric flight test, arguing that from such a height he might drift out of British air space and land in environs where the suit could be captured by unfriendly forces.

Ridge returned to the United States and worked with the Liquid Carbonic Corporation in his native Massachusetts on the design of a thermal pressure suit. But when the company discovered problems with the system's oxygen mask shortly before the first flight test, subsequent tests—along with the entire engineering program—were canceled due to fears of bad publicity in the event that Ridge was killed or injured at altitude.

Ridge next went to the Army Air Corps with his own pressure-suit design and proposed a cooperative venture, but failed to convince the brass of the practicality of high-altitude survival suits. In the meantime, a British pilot succeeded in reaching the stratosphere in 1936 wearing a variant of the Haldane-Davis suit Ridge had tested in the chamber. As the seeds of world war germinated in Europe, the U.S. Army Air Corps changed its mind and moved aggressively into pressure-suit design and fabrication. But Mark Ridge had by this point faded from the picture. Sadly, Ridge spent much of the rest of his life in a series of mental institutions.

The legendary Wiley Post had more success. Post, the former stunt flyer, air racer, and commercial pilot who made the first solo flight around the world, had himself identified the need for a high-altitude flying suit in the early 1930s, understanding that a pressurized-oxygen breathing system would be a crucial component of any successful design. With help from James "Jimmy" Doolittle, Post convinced B. F. Goodrich to build him a pressure suit. Russell Colley built the suits for Goodrich that Post wore on his maiden stratospheric test flight in 1934 and to 50,000 feet the following year, earning Colley the designation of Father of the American Space Suit.

The biggest problem with these early suits was the immobility they created at full inflation. Their constriction simply didn't allow aviators to perform the basic tasks required of an active pilot in the cockpit, which was one reason why full-pressure suits were not issued to American flyers during World War II.

The Dave Clark Special

By mid-January 1965, while the weather kept the Lakewood parachutists mostly on the ground, things appeared to be looking up for Nick and his project. He had found some sympathetic ears at the David Clark Company, which had been in the business of air-crew protection systems since 1941 and which manufactured partial- and full-pressure suits for the Air Force and for NASA. Nick had immediately driven to Worcester, Massachusetts, and had gotten a tour of the David Clark plant. He met with company president John Flagg and with David Clark's Gemini spacesuit program manager, George Johnson, a retired USAF physiologist. Flagg didn't agree to sell Nick a suit, but he did have him measured and, with Johnson, openly discussed ideas for modifications they felt would be in order for a super-high-altitude free fall. These modifications included a heated helmet visor, a fitting for electrical power to heat undergarments, and the removal of a pressure-sealing zipper to further reduce the suit's already minuscule leak rate. Johnson provided names of contacts at the Air Force who might be able to arrange for pressure-suit training at Air Force facilities, possibly even at nearby Otis Air Force Base on Cape Cod. It was all very encouraging. Nick asked Senator Williams to write to Dr. Smith Ames at the Air Force Surgeon General's office in Washington, D.C., to inform him of the nature of the project and to request pressure-suit training.

Dr. Ames was interested in what he heard and promptly got in touch with Nick. However, he explained, any civilian request for Air Force training would have to be made through John Sims at the FAA. It looked like another stonewall.

Nick spent February 3 in Washington, D.C., in further meetings with NAA officials who had agreed to arrange for barographs to record what everyone seemed to hope would be a record altitude. He dropped by Williams's office, but the senator was out of town. In conversation with the staff, Nick asked whether Williams would object to his name being used for fundraising purposes. The project was still short on money and a prestigious name from the political world might bestow some additional legitimacy on the whole enterprise. The senator later explained why he could not allow his name to be used in such a way but recommended that Nick approach some military and corporate contacts Williams would provide.

Finally, later that month, Senator Williams had some good news. He had gotten the attention of John Sims at the FAA, and Sims's imagination had been "sparked"—to use Sims's own word—by the plans for a world-record parachute jump. Sims wanted Nick to drop by for a chat on his next trip to Washington and asked whether the Don Piccard who was listed as the balloon builder for Nick's project was, in fact, a member of the "famous Piccard family." He assured both Williams and Nick that the FAA and the Air Force would be quite happy to cooperate by providing training—assuming, of course, that Nick could manage to acquire a pressure suit elsewhere. The elements were *almost* in place now: the Air Force would provide the necessary training if Nick could find a suit, and the David Clark Company would provide the suit as long as the Air Force agreed to provide training.

In 1962 the Aeromedical Systems Division of the U.S. Air Force had issued some fairly specific directives to the David Clark Company on the control of its pressure-suit technologies. The biggest concern about the suits at that time had been to "preclude their acquisition by any unfriendly power," and sales to foreign governments would always require formal USAF clearance. At the same time, the Air Force didn't categorically prohibit qualified domestic companies from acquiring suits as long as sufficient investigation into their intended use had satisfied all involved and as long as the appropriate written approvals and clearances had been obtained. None of the directives anticipated a civilian individual's attempt to acquire one of the suits.

On February 15 Nick arrived in Worcester by invitation for a pressure-suit fitting at David Clark and was flabbergasted to learn that the company had apparently decided to build a custom version of one of its highly successful Air Force suits—the A/P22S-2, known within the company as the "Dash Two"—specifically for Nick. David Clark had built some five hundred of these suits over the previous six years.

The David Clark Company occupied a weathered nineteenth-century brick building on a hillside just beyond downtown Worcester that had originally housed a leather processing and manufacturing plant. Clark's engineers had begun working in the mid-1940s on antigravity suits designed to prevent the rapid shifting of bodily fluids as a result of in-flight G-forces. In 1955 the Air Force's Air Research and Development Command ordered the Aeromedical Field Laboratory at Wright Field to develop a pressurized garment for jet pilots that would provide sufficient mobility even at full inflation. More than two dozen contractors submitted bids, and one of the designs adopted was David Clark's XMC-2-DC. A refined version—the A/P22S-2, formally referred to in the company's literature as a "High-Altitude, Full Pressure, Flying Outfit"—was delivered a year later at a cost of $3,891 per unit.

In addition to the four-layer coverall itself, the assembly included pressure socks and exterior boots, gloves, a suit-pressure controller, a demand oxygen regulator, and a full-pressure helmet. An aneroid device was designed to inflate the suit automatically at an altitude of about 35,000 feet. In high-altitude operation, the suit was pressurized to the equivalent of between 34,600 and 36,200 feet.

The breakthrough design of the Dash Two included a fishnetlike layer of nylon called "linknet," which prevented the suit from ballooning and severely restricting mobility, as well as improved helmet-visor defogging and a general increase in overall pilot comfort. It wasn't perfect: in spite of mobility improvements, pilots would always demand more freedom of movement. In addition, the suit still had a tendency to leak a bit more than was ideal. But it was the first practical full-pressure suit for pilots and it subsequently became standard-issue for certain classes of Air Force jet pilots. The Dash Two was the precursor of the Gemini astronaut spacesuit. Scott Crossfield wore a similar suit on the first X-15 flight in 1959, as did Neil Armstrong on his X-15 flights two years later.

It was this same suit that David Clark—in an unprecedented move—was now modifying for a civilian: Nick Piantanida. The company had a facility fully equipped and staffed for this purpose on its top floor. The process of donning a pressure suit is elaborate, and while it was possible to get into the David Clark suit without assistance, one or two technicians generally assisted. One of the problems with the fit of the suits from Strato-Jump's perspective was that the body positions assumed by a skydiver are significantly different from those required of a jet pilot, and the suit would need to accommodate these rather unorthodox gyrations.

The people at David Clark had appreciated Nick's professionalism (he always arrived promptly, was always well dressed, and always carried his briefcase) as well as his desire, but they were flat-out awed by his physical strength as they watched him pull his limbs into a skydiver's arch posture while wearing a tight-fitting suit of nonextensible material that in theory made such a maneuver impossible. The suit, they agreed, would require modification to allow for Nick's abnormally large shoulders and to make it easier to manipulate. There would also be some thermal issues to be resolved, but those could come later.

Jack Bassick, who would in time become the company's director of research and development, remembers Nick's early visits to Worcester: "He seemed like an impressive, almost larger-than-life kind of guy who was determined to do what he wanted to do. A big, strong guy. He had these huge shoulders and this tiny waist. And because we had to customize the suit for him and the suit had to mimic his shape, it turned out to be very different from virtually any other suit design we had. He was a very, very strong and powerful guy—and probably strong-willed as well."

On the euphoric drive back to Brick Town, Nick found himself wondering what the catch was. At the same time, he told himself that there was no way that a prestigious firm like David Clark would be going to the expense of manufacturing a custom pressure suit if they hadn't already concluded that they were going to make it available to him when the time came. That night, cautiously triumphant, he wrote in his log: "Finally!! But, I don't have it in my possession yet."

One week after the exhilarating trip to Massachusetts, Nick and two of his Lakewood parachute buddies got themselves arrested at a carnival, a benefit for a children's hospital, in White Lake, New York. And that wasn't the worst thing that happened to Nick that Sunday.

The three jumpers had intentionally set out to violate a thirty-eight-year-old New York state prohibition against exhibition jumps in hopes of having the opportunity to challenge the law's constitutionality in court. They got it. Less than a month later the charges were dismissed and parachute prosecutions under section 245, subdivision 13, of the General Business Law effectively ended in the state of New York.

The real problem for Nick had occurred when he exited the airplane at 5,000 feet above White Lake. His chute—for the first time since he'd begun jumping a year and a half earlier—deployed prematurely, slipping out of its pack as Nick left the plane, almost unquestionably the result of his own mistake in packing. For a bone-chilling moment he feared that the lines might snag the plane's landing gear. But he landed safely with his rip cord still intact. He put a good face on for the police who met him but later admitted to a Union City reporter that he'd been shaken by the incident. It was a sober reminder that parachuting was still a very risky business, regardless of the jumper's experience or skill. One lapse of concentration, in the sky or while packing the chute, and even the most skilled skydiver in the world could quickly become another "ground impact" statistic.

The following month Nick attended a week-long Aerospace Medical Convention at the invitation of Don Nesbitt, president of Buffalo, New York's Firewel Company, the manufacturer of the David Clark Company's pressure-suit oxygen breathing system. Nesbitt wanted to introduce Nick to people who might be able to help him—specifically Maj. Harry Collins and Dr. Smith Ames of the Air Force—and he wanted to hear their opinions of Project Strato-Jump firsthand. It was a great opportunity for Nick to sell himself and his dream, and by all accounts he took good advantage of it. He returned home with a tentative commitment from Firewel to donate all of the oxygen-system equipment for the project, as well as the blessings of Collins and Ames, who promised to argue Nick's case in influential circles.

On May 3 Nick returned to Worcester for final adjustments to the custom suit with the expectation that he would take it back to New Jersey with him for the first time. When he arrived at the David Clark plant, however, he learned that work on the suit had stopped on the previous Friday when the company had received a memo from Air Force Systems Command at Wright-Patterson Air Force Base listing a number of provisions that would need to be satisfied before the Air Force could allow the suit to be released into Nick's custody. The people at David Clark liked Nick and wanted to

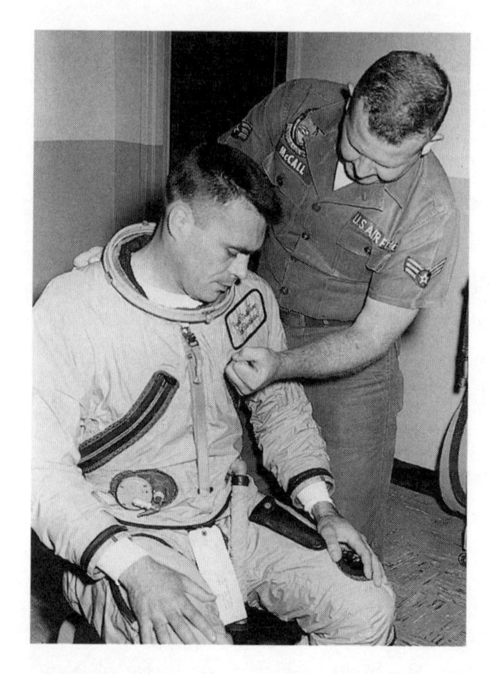

Marvin McCall assists Nick with the Dave Clark Special at Tyndall Air Force Base. (Courtesy USAF)

help him, but the prime consideration for the business was, of course, protecting its status as a preferred Air Force contractor.

Nick immediately got on the phone with a colonel at the Air Force Surgeon General's office who had already been briefed—probably by Smith Ames—and who in turn placed a call to Wright-Patterson and got the suit released to Nick, at least temporarily. Technicians completed work on the bright orange garment with a special oval breast-patch insignia bearing the legend, "Nick Piantanida: Project Strato-Jump," and turned it over to Nick, who rushed it triumphantly back to Brick Town. That night he squeezed himself into the suit and modeled it for Janice who regarded it with mixed emotions.

"Can you believe I've got this?" he asked. "You just don't understand what this is. Look at this!"

She was genuinely happy for Nick, but for the first time she felt a distinctly troubling shudder of anticipation for what might lie ahead for them both. "I would never have said it to him, of course," she admits, "but I was hoping he wouldn't get the suit."

Now that he had it, dominoes began to fall. Nick received a call from a Capt.

Marion Pruitt, officer in charge of the Physiological Training Unit at Tyndall Air Force Base in Panama City, Florida, who explained that it was his unit's intention to make its training facilities available to Nick at his convenience. Whatever they could do, Nick should not hesitate to ask. The near-overnight change in attitude was almost eerie. In a letter confirming the training arrangements, Pruitt wrote, "Your résumé is quite impressive. I do not know where a better qualified individual could be found to attempt this undertaking. We in this organization are delighted to have the opportunity to assist in any way that we might with this important project." Training was promptly scheduled for May 12, a little more than a week away.

Nick picked up an unemployment check in Union City on May 11 and immediately left for Florida by car. He arrived the following night, having asked Pruitt and staff to reschedule him for May 14, Friday. Upon arriving in Panama City, Nick found a discouraging message from the David Clark Company waiting for him. The gist was that the formal Air Force approval had not yet been granted and that Nick was requested to return the suit without delay. A phone call to Smith Ames in Washington helped secure permission for Nick to complete his training at Tyndall using the suit, with the understanding that he would deliver it back to David Clark as soon afterwards as practical.

On May 14, with Capt. William Nicks supervising, Nick Piantanida became the first civilian in history to undergo formal pressure-suit indoctrination by the U.S. Air Force. He also participated in elements of Tyndall's ejection-seat training program. Nick made two simulated high-altitude flights in the big decompression chamber using the David Clark suit and the Firewel oxygen apparatus. The first was a forty-five-minute "flight" to 70,000 feet. Technicians had explained the functions and operation of the suit and the oxygen system to Nick and were able to monitor him constantly during the rapid decompression. The second chamber run, to an 80,000-foot equivalent, lasted for one hour and twenty-five minutes. Nick received a rating of "satisfactory" on his performance.

Before both chamber runs, Nick was given 100 percent oxygen to breathe for fifteen minutes. This process, known as *prebreathing*, had become a standard component of preparation for high-altitude flight and had been used religiously in all the military stratospheric balloon programs. It was designed to purge nitrogen bubbles from the bloodstream and body tissues to avoid their expansion with the decrease in pressure, thereby preventing the disability associated with decompression sickness and, specifically, a painful phenomenon known to deep-sea divers as "the bends." On the first run Nick had also been instructed to briefly open his helmet visor at 25,000 and again at 35,000 feet in order to experience and familiarize himself with the debilitating symptoms of hypoxia.

Both of these procedures—prebreathing of 100 percent oxygen and the intentional opening of the visor—would become major issues a year later. In the spring of 1965, however, it was all part and parcel of what seemed nothing less than a mounting triumph for Project Strato-Jump.

As he sped back up the East Coast following the training, Nick marveled at the professionalism and enthusiasm with which he'd been greeted at Tyndall. It was a complete 180-degree turnaround from the reception he had become accustomed to from the military. He felt like the men he had met in Florida understood not only what he wanted to do, but why. The whole team had fallen all over themselves trying to help. He had even talked to a couple of the technicians about serving on the support crew for Strato-Jump. What's more, Captain Nicks had informed him that, having completed the Tyndall course, he would now be received with open arms at the FAA facilities in Oklahoma City—which included a cold chamber and high-altitude aircraft for jump training—should he be interested in scheduling time there.

In spite of the depressing news that he'd have to return the pressure suit to David Clark, it had on balance been a very good week.

As an on-again/off-again driver for Smith Transport and a member of the Teamsters Union (Local 560), Nick had made inquiries early in 1965 concerning help with fundraising for Strato-Jump. He began by asking Sam Provenzano to try to convince his brother Anthony, president of the Union City Local and vice-president of the International Brotherhood, to agree to serve as committee chairman—or, if not as chairman, as trustee—for a fund drive. Tony Provenzano dismissed the idea citing legal restrictions, but agreed to consider appealing to Jimmy Hoffa for the financial support of the International. As a last resort, Nick tried a mass mailing to the membership of the Teamsters Local and to the local Shop Stewards Union where he also had contacts, but calculated that he spent more money on envelopes and postage than he received in donations from the drivers or shop personnel. So much for the Brotherhood. In the end, all Nick really got from the Teamsters was a prominent story playing up the competition-with-Russia angle in the *Hudson Dispatch* placed by a reporter for the local Teamster house organ.

The lack of assistance from Local 560 officials was surely, however, a blessing in disguise. The Piantanida boys had grown up across Third Street from Teamster insider Jimmy "The Hook" Cirillo (who got them their union cards so that they could drive trucks when they needed money, in spite of the fact that Nick always refused to pay the required dues), and they knew that these guys were thugs. What they didn't know at the time was what *kind* of thugs. Union City's Local 560 was one of the most notoriously corrupt labor unions in American history, and "Tough Tony Pro" ruled the organization through pure terror. He was twice convicted of extortion in the late 1950s and was later sentenced to life in prison for the murder of the Local's secretary-treasurer, Tony Castellito, who had had the temerity to challenge Provenzano for the presidency. Tony Pro was even convicted on RICO charges for running an extortion racket from his prison cell. When these guys did you a favor, you were in a perpetual—and precarious—state of debt. Project Strato-Jump had enough problems.

Efforts to raise money—legitimate money—would continue throughout 1965. Nick was determined to leave no stone unturned. A local priest at St. Anthony's in Union City agreed to serve as a project trustee in the hopes of pulling in some Catholic money. A friend of a friend donated $500 in February and brought in an additional $250 from a colleague. Shop owners in the old neighborhood did what they could; Carmine's Beauty Salon in Union City, for example, chipped in $25. At the suggestion of Senator Williams, Nick approached executives at Johnson and Johnson Corporation and got a commitment from them to consider a donation, and followed that up with a trip to Wall Street, where he called on a number of Jacques Istel's high-rolling financial associates to discuss corporate contributions. He even wrote to Adm. Hyman Rickover at the Atomic Energy Commission for permission to use his name in fundraising appeals, a request that was predictably denied.

In April Nick decided to go straight to the very top and wrote to both President Lyndon Johnson—whose 1958 Senate Preparedness Committee had raised concerns about Russian advances in space technology to new heights and who, following the successful launch of the Soviet satellite *Sputnik*, had publicly compared America's situation to both Pearl Harbor and the Alamo—and to the First Lady at the White House. In explaining his project, Nick wisely placed particular emphasis on the importance of capturing the world record from the Russians and of asserting U.S. dominance in space and asked for the First Couple's official blessing. He received a polite note from Ladybird's social secretary and another from one of the president's aides saying that Nick's letter had been forwarded to the FAA for consideration.

But the most promising leads at this point were coming courtesy of Barry Mahon who had been putting out feelers in the television broadcast industry. In March RKO-TV indicated that it might be willing to commit as much as $10,000 for exclusive film rights to Strato-Jump. By May the offer had dropped to $5,000, and Nick was beginning to balk. Istel had consistently urged him not to sign away anything for a lowball figure, and if the project turned out to be successful, $5,000 just didn't seem like enough. The estimate at that point was that the entire venture would require a total budget in the neighborhood of $75,000 (the actual cost turned out to be quite a bit higher). Nick had reason to believe that he could offset as much as $50,000 of that with donations of equipment and services. But the balance still looked formidable, and RKO-TV's offer just wouldn't make enough of a dent.

While Nick was scheming ways to bring in the rest of the $25,000, he barely had enough cash on hand to pay his monthly rent. In the tax year 1964, Nick had earned a total of $3,898 in truckdriver's wages, only a little more than half the average annual income for Americans at that time. He was regularly forced to borrow small amounts to pay for trips and jump fees. All of his meager savings and most of his energy were being funneled directly into the dream. But if it all came together and Nick was able to break the record and establish himself as a bona-fide American hero, he was convinced that with his promotional savvy he could parlay that status into a comfortable living for his family.

He never once hedged that bet. There were no backup plans, no contingencies beyond Strato-Jump.

Following his training at Tyndall, which had been administered by both USAF and FAA personnel, Nick wrote to each individual on the pressure-suit indoctrination team, thanking them for their assistance, expertise, and dedication. In one such letter to Dr. Stanley Mohler, director of the Civil Aeromedical Research Institute of the FAA, he wrote, rather over-optimistically, that as a result of his time at Tyndall he had gained a "complete understanding of the problems" involved in Project Strato-Jump. In another letter to Air Force headquarters, Nick singled out pressure-suit technician Marvin McCall whom he felt had exhibited an admirable zeal. He also sent words of gratitude and praise to John Sims at the FAA whose interest in Strato-Jump had opened doors into not only his own agency but into the Air Force as well, and to Maj. Gen. Robert Bohanon (USAF) for issuing final approval for the training at Air Force facilities.

As soon as Nick had returned to New Jersey following the training, he was contacted by David Clark officials who reminded him that, due to the absence of formal Air Force clearance, the suit would have to be returned to them no later than May 25, eight days later. In spite of a hectic schedule during those eight days, which included a number of press interviews, one with ABC-TV, Nick stayed in close touch with Senator Williams, who continued to plead with the assistant secretary of defense to intercede with the Air Force. Williams was assured that an approval would be issued to the David Clark Company by May 24 and immediately forwarded the news to Nick by telegram.

Sadly for Nick, no such approval ever showed up in Worcester and on May 25 he was forced to once again make the drive to the David Clark plant to return the pressure suit. Company officials were apologetic but firm. They would not buck the Air Force on Nick Piantanida's or any other civilian's behalf. While in Worcester, Nick talked with Senator Williams by phone and discovered that the Air Force's stated objections to releasing the suit centered on the Dash Two's heating capacity, which they judged to be inadequate for a stratospheric free fall. The Air Force also insisted that the suit was generally unsafe under conditions of sustained pressurization, and that there was a serious risk of "incapacitation" if no stabilization device such as a drogue chute was employed during the long free fall.

As soon as he returned home, Nick gathered the data from the testing at Tyndall and forwarded it to the Air Force, hoping it would at the very least alleviate concerns about performance under sustained pressurization. He also suggested that the modified suit be renamed—"A/P22S-2" was a special Air Force designation—so as to remove any possible association with the Air Force and to eliminate any liability concerns for the Department of Defense.

On May 27 Senator Williams—having finally reached the end of his patience—wrote a long and pointed letter to Air Force Lt. Col. James Larkins of the Office of Legislative Liaison at the Pentagon in which he provided a summary of the bureaucratic maze to which the organization had subjected Project Strato-Jump. In the letter Williams made it clear that inquiries initiated by his staff had led him to conclude that the Air Force had intentionally and selfishly stonewalled Piantanida because it was reluctant to help a civilian break one of its own most cherished aeronautical records, Joe Kittinger's unofficial world free-fall mark. Williams argued that this consideration should properly be irrelevant to the Air Force's decision to release the David Clark Company's suit, a release Williams claimed—despite the company's insistence on obtaining it—should not even be required.

"As I see it," Williams wrote, "the problem is as follows: My constituent . . . would like the Air Force to either say a release is not necessary (a true fact) or grant the release (unnecessary, but thereby continuing the recently established good relationship between the Air Force and the project)." Williams's letter concluded: "Like any project of this size, a delay in a decision similar to the one we are presently experiencing, can throw the whole operation out of kilter for a long enough period of time to destroy the original operating schedule, thereby destroying the investments of many individuals."

Nick's obsession was apparently becoming the senator's obsession as well, and the latter was something the Air Force brass clearly did not relish. Exactly two weeks later Nick received a letter from Williams confirming the Air Force's formal release of the David Clark suit (now renamed "The Dave Clark Special," an idea of Nick's that appealed to all concerned and may even have helped Williams close the deal).

"I am sorry," Williams told Nick, "that the letter was so long in coming. The whole affair took on the guise of a comic opera for a while and I think the Air Force is truly embarrassed, though an admission of this fact would never be forth coming." A few days later a member of the senator's staff forwarded the official paperwork to Nick with a note saying that Williams had simply grown "tired of their foolishness" and expressing hope that the matter was finally resolved for good.

Donning helmet at FAA facility in Oklahoma City prior to cold-chamber session. (Courtesy USAF)

In a follow-up letter to Williams, the Air Force announced its formal disassociation from Nick's project and from his use of the pressure suit. The letter reiterated the earlier warning that, based on its own high-altitude tests, Nick would be exposing himself to dangerous, perhaps deadly, aerodynamic forces unless a drogue chute or some other stabilization device was employed.

After signing a liability release in Worcester and taking temporary possession of the Dave Clark Special, Nick was again on his way, traveling to Oklahoma City in late June to train at the FAA Aeromedical Center there. The Oklahoma site offered a cold chamber that was necessary to test both the suit and the oxygen system at the temperatures to which they would be exposed in the stratosphere. The FAA was also well equipped to conduct physiological training for super-high-altitude pilots and would help determine how effective the suit's thermal properties would really be in the extreme conditions Nick would face and to measure how well his body would tolerate those conditions.

On June 25 Nick was dressed in heavy cotton socks, heavy waffle-weave thermal underwear, exposure mitts over the pressure gloves, and the David Clark pressure suit. He was escorted into the cold chamber at -73 degrees Fahrenheit where he remained for seventeen minutes while a fan circulated air inside the chamber at roughly fifteen miles per hour. Skin temperatures were measured with thermocouples taped to twenty-one different spots on Nick's body. It was important to determine which areas would suffer most from the severe cold during the free fall to Earth so that they could receive extra protection. The lowest temperatures discovered on Nick were located on his insteps, a result some suggested of his standing on a bare concrete floor during the test. Nick's subjective report was that while he had been fairly comfortable during the test, he felt the cold most acutely on the insteps, on his thighs, and on the tip of his nose. His core body temperature had dropped a fraction of a degree, but everything was within acceptable limits.

The FAA observer supervising the cold-chamber run concluded that Nick should have no serious problem with low temperatures during the free fall, especially since on the real jump his equipment would be augmented by electrically heated gloves and socks, an electrically heated visor, and a coverall to act as a windbreak.

A Balloon of One's Own

Nick had made a quick trip to Minneapolis earlier that month to meet Don Piccard for discussions about the balloon that Strato-Jump needed. It was the first face-to-face meeting between Nick and Don, and it appears to have gone better than anyone could have expected. Nick had initially been a bit put off by Piccard's aloof, somewhat imperious bearing, but he was heartened by the extent of Piccard's knowledge of balloon-building techniques and materials, and by his extensive network of useful contacts. Piccard found himself pleasantly surprised by Nick Piantanida in the flesh.

"Perfectly proportioned," Piccard recalls, "a physical ideal. I had great confidence in his whole attitude, his thoroughness, his qualifications. I think I may have been a little concerned by the accent at first. But he proved that he wasn't some dumb taxi driver from Brooklyn, which is the stereotype you might get from his accent. Like a Limey instead of an Oxford person, you see. Maybe he didn't have the formal stature to wander in the Elysian Halls of academia. But he was, I think, very smart. Probably too smart for his own good."

Nick had needed to borrow $200 to cover his trip expenses and, once in Minneapolis, had wired Istel for additional funds when he realized Piccard—who was now based on the West Coast—expected Nick to pick up his expenses as well. Piccard, who with his combustible temper and predilection to sudden rants was spectacularly ill-suited to corporate life, had recently been fired from Raven Industries and had relocated to Costa Mesa, California, to start his own sport-balloon company. "I was never a team-type person," Piccard confirms. Nevertheless, he had offered his services as a consultant to help Nick select an established contractor for the project.

Piccard and Piantanida visited the Applied Science Division of Litton Industries in St. Paul, which had only months before acquired the very capable General Mills balloon team; Piccard's father had been an important player at General Mills and Don knew the principal engineers at Litton. They traveled thirty miles south to Northfield, Minnesota, to tour the facilities at the G. T. Schjeldahl Company where Piccard had worked briefly prior to his stint at Raven. Finally, they headed for Sioux Falls and held discussions with Raven Industries. The one facility they avoided was Winzen Research in Bloomington, Minnesota. Not only had Otto Winzen ignored earlier requests for information (unbeknownst to Nick, Winzen was vehemently opposed to civilian record-chasing and, to some extent, to recreational ballooning), but Don Piccard's father and Winzen had had a serious rift many years before at General Mills that precluded any possibility of a working relationship between Otto and Don.

Piccard made no secret of the fact that he favored Schjeldahl, which had an excellent track record with its high-altitude scientific balloons, for Project Strato-Jump. These balloons were quite different from those available from Litton and Raven (and Winzen as well). Schjeldahl produced Mylar envelopes, and Piccard made the point that reinforced Mylar—which is much stronger than polyethylene—would offer a demonstrably greater safety factor. In spite of the fact that a Mylar balloon would be considerably more expensive, Piccard reasoned that it would provide the best possible chance of success and urged Nick to consider it. Since Nick had felt compelled to dispense with his original notion of a staggered series of flights to higher and higher altitudes, it was understood that Strato-Jump would have to go for broke. This meant that there would be no chance to experiment with envelope designs and materials: the balloon would have to work the first time. Piccard's case for Mylar was a strong one.

Upon returning to New Jersey, Nick followed up the Midwest trip by contacting both Litton and Schjeldahl and asking them to submit bids for a Strato-Jump balloon. Piccard, meanwhile, contacted both companies himself and urged them to cooperate with Nick.

Litton Industries had brought with it from General Mills an important contract with the Air Force Cambridge Research Laboratory (AFCRL). The original purpose of the plastic balloons built for the AFCRL was to carry atmospheric-sensing apparatus aloft to monitor radioactivity from atomic-bomb blasts around the globe, and Litton was continuing the launches that had begun at General Mills. Karl Stefan, who ran the scientific balloon operation, had gone to work in the wind-tunnel laboratory at the University of Minnesota after retiring from the Navy. When the university lost much of its funding for the wind tunnels, Stefan accepted an offer from General Mills to conduct particle physics research. He inherited the balloon operation primarily on the strength of his knowledge of sport ballooning, much of which came second-hand through Stefan's wife, Lucy, who had been active in recreational gas ballooning with Don Piccard and others.

By the time Nick approached Litton Industries in the summer of 1965, the balloon team was struggling to stay afloat. Defectors from General Mills had created powerful competition in the form of Winzen Research and Raven Industries, and the AFCRL contract was about all that was keeping Litton in the game. So, in the face of serious doubts on the part of Litton executives in California and a general nervousness caused by the recent poor performance of the company's stock, Karl Stefan jumped at the chance to build and launch a high-altitude balloon for Project Strato-Jump, despite the fact that Stefan had some reservations of his own.

"I was very suspicious at first," Stefan admits of his initial meeting with Nick and Jacques Istel. "I thought, here's a couple of kooks. But then they began to show that they had done some homework. We just got along real well. So I developed some confidence in them, enough to go ahead with the project."

Stefan felt that the potential for positive publicity from a high-profile manned flight was the Litton group's best chance for survival in the marketplace, and he was willing to risk the company's—and his own—credibility to make it happen.

On June 29, just four days after Nick's FAA training in Oklahoma, Litton Industries submitted a formal proposal and a time-and-materials bid for the balloon. The bid specified that Strato-Jump would be liable for cost overruns but would receive a refund if the actual costs came in lower than the estimate. Litton also called attention to the fact that if any equipment belonging to the company was lost or damaged during the flight, the client would be expected to cover the expense.

The balloon being proposed was a 3,690,000-cubic-foot, 1-mil (1/1,000th of an inch—or .0254 millimeter—in thickness) polyethylene envelope with Fortisan filament load tapes capable of raising a 700-pound payload to a height of 118,000 feet. Polyethylene resin, first synthesized in 1933 through a chemical reaction between ethylene and benzaldehyde, exhibited superior performance at stratospheric temperatures and was relatively easy and relatively inexpensive to fabricate; Fortisan was a synthetic fiber in the rayon family with extremely low elasticity. The Litton group had built a number of similar balloons for the Air Force and had a high degree of confidence in them. The balloon itself would weigh about 745 pounds and would have a surface area of approximately 120,000 square feet, nearly three acres of plastic. But because Nick intended only a one-way trip in the balloon, there would be no descent valve like on the Air Force's Manhigh system. The gondola's only means of return to Earth would be beneath the big forty-eight-foot parachute attached to the top of the frame.

The proposal for the gondola, an unpressurized enclosure nearer in concept and design to the Navy's *Strato-Lab V* craft than to any of those used by the Air Force, specified an aluminum framework, welded at the joints—46 inches wide, 46 inches long, and 54 inches high—and covered by a protective shell of two-inch-thick polystyrene. This would provide enough space for a large man in a fully inflated pressure suit to share the confines with the required instrumentation and auxiliary equipment. The front of the rectangular compartment would provide a six-foot, two-inch vertical clearance when opened, sufficient to allow Nick to stand comfortably and to leap out when the time came.

The proposal itemized the anticipated costs (see table on page 78). It also itemized the Litton equipment for which the project would be charged in

Litton Industries' Quote for Equipment and Services: Strato-Jump I

Operations	
labor	$6,428
materials	$602
	$7,030
Equipment	
labor	$2,157
balloon	$5,055
gondola	$1,180
	$8,392
Miscellaneous (helium, transport, meteorological)	$2,504
Total	**$17,926**

the event of loss or damage during the flight. The dollar figure for that equipment amounted to another $2,562.

For the first time, Project Strato-Jump and SPACE, Inc. had a good idea of what a world-record free-fall attempt was going to cost. The Litton proposal did not, of course, include the expense of a pressure suit, oxygen system, or parachute hardware. But Nick had suppliers for those items and had a reasonable expectation that all could be composed of borrowed or donated equipment. He had also reached an agreement with Athenia Aircraft Supply of Passaic, New Jersey, which had agreed to donate additional miscellaneous equipment the project would need.

On July 2 Nick informed Don Piccard that he intended to accept the Litton proposal. He had received no response from Schjeldahl and was impatient to move forward. Piccard continued to argue the merits of reinforced

Mylar balloons for manned stratospheric flight and offered to contact Schjeldahl again on Nick's behalf. There were indications that Schjeldahl had simply ignored Nick's original request for a quote, suspecting that he didn't really have the financing to launch his project.

Nick, however, was in no mood to wait. A few weeks before, in mid-June, he had drafted a general statement authorizing Jacques Istel to act as business director and to negotiate on behalf of Strato-Jump. And, after clearing the deal with Istel, Nick wrote to Karl Stefan at Litton on July 6 accepting their bid in principle and inquiring about contract arrangements. Having made his decision, Nick was determined to put the project on the fast track. A proposal from Cliff Merrell at G. T. Schjeldahl finally arrived in Nick's mailbox two weeks later. It quoted a Mylar envelope the size of the one being proposed by Litton at $16,000, a figure that included neither operations nor a gondola.

Nick, eager to see balloon and gondola construction get underway before anything went wrong or anyone changed his mind, wrote to Thomas (Tim) O'Malley—Karl Stefan's second-in-command—asking for a schedule of payment terms. It now appeared likely that he would get his trip into the stratosphere before the end of the year. But, optimism temporarily aside, he knew from experience that things were not always as they seemed. A million things could go wrong, most of them out of his control. The faster he could nudge the project toward launch of the balloon, the faster he could seize control of his own fate. All the rest of them needed only to get the balloon up to altitude—Nick would do the rest.

Which is not to suggest that his training regimen was complete. He still needed some actual jump testing of the fully inflated pressure suit with Pioneer Parachutes' custom high-altitude rig, something Nick had been unable to accomplish so far because of the difficulties in getting custody of the suit for any length of time. He had made dozens of jumps at Lakewood without the suit (or with a noninflated suit) in an effort to work out the details of the parachute system. He had consulted with Pioneer engineers on the design of a special harness and on release mechanisms that would provide maximum mobility without compromising safety, and tests had shown that the system would allow the jumper to free fall in either a face-to-earth or back-to-earth position.

Pioneer's system contained three parachutes. The modified twenty-four-foot Para-Commander was the main canopy and would be carried in Nick's backpack. The reserve, a modified military-type C-9 canopy with a T-slot cut at the rear to provide greater maneuverability, was carried in a bulky chest pack. And finally, a six-foot hemisflow drogue chute would be stowed in a tandem arrangement with the main canopy in the backpack even though Nick had no intention of using the drogue. To employ his skydiving skills to demonstrate that a stabilization device was unnecessary for a trained jumper—even in the stratosphere—had become one of Strato-Jump's primary objectives. Drogues, Nick would say, were only "so much useless bulk." Nevertheless, Pioneer prudently argued for the inclusion of the drogue in the event that Nick found himself unable to maintain stability in the extreme conditions in the stratosphere.

Both the main and the drogue were equipped with quick-release Capewell fittings which would allow them to be jettisoned instantly in the event that the reserve had to be deployed. Both the main and the reserve chutes were attached to barometrically controlled automatic-opening devices provided by the U.S. Gauge Company. These devices would release the canopies at predetermined altitudes (6,500 to 8,500 feet for the main, around 4,000 for the reserve), although Nick would always have the option of pulling a ripcord and releasing the main before the aneroid popped the pins. The reliability of these automatic openers had already been proven in repeated jump tests at Lakewood.

Nick was active at Lakewood during the first three weeks of August. He jumped three times on August 1, made his first jump with a new Para-Commander three days later, and participated in the Lakewood Open on August 7, in which he took second place in the accuracy competition. Then, to complete his Strato-Jump parachute training, Nick made eleven more jumps at Lakewood between August 12 and 20, each with the pressure suit fully inflated. Five of the jumps were from above 10,000 feet and, as Nick had suspected, the inflated suit forced him to make major aerodynamic adjustments in order to control his attitude and to avoid tumbling or spinning. He even deployed the drogue on a couple of the jumps to make sure it would release and perform as expected in the event—unlikely in Nick's mind—that it would be needed.

On the first test jump in the pressurized suit, Nick left the airplane at 15,200 feet. "The slightest arm movement caused a wavering from side to side," he wrote later in his training log. "Weird feeling. No direct pressure from ambient. Config causes slight head down position. Arms cause stability change with little effort. Fall felt very unsteady." On a subsequent jump two days later, Nick found that he was able to eliminate the wavering problem by exaggerating his arch position, craning his neck back and cocking his legs slightly behind him, and was able to execute controlled 360-degree turns. A follow-up on August 16 was the highest of the pressurized test jumps at 17,500 feet. By this point Nick was reporting no stability problems whatsoever. He had made the necessary adjustments to his technique and was skydiving with precision in spite of the heavy, bulky suit.

Accelerating flat-spin had always been the high-altitude parachutist's most-feared mechanical force. Nick had described the problem in layman's terms to a reporter earlier that year: "If the spin of your body increases from 80 revolutions to 120, your brain becomes dislodged, your internal organs get all mixed up, you fail to function and you're dead." On his final pressurized jump, Nick deliberately initiated two high-speed flat spins and found that he was able to arrest both spins within one second. He experimented with various skydiving body positions with names like the Half Delta and the Cannorozza. He tried radically shifting his center of gravity and even flailing his arms and legs in a "swimming" motion. In the end, the most effective position was one known as the Full Delta, an arms-back, head-to-earth configuration in which maneuvering is achieved by subtle movements of the shoulders, hands, and legs. An in-air observer reported that on one of the spins Nick was twirling so fast that his arms and legs began to blur, but that the spin was arrested almost instantly once Nick pulled himself into the Full Delta. It now looked like a good bet that Nick's skill would give him the ability to handle himself in free fall in spite of the aerodynamic handicap of a fully pressurized suit. He had never doubted it, but now others connected with the project could be assured that, based on the FAA testing, Nick wouldn't freeze and, based on the pressurized jump trials at Lakewood, he wouldn't black out in a flat spin. All indications were—in spite of the protestations of the Air Force and others—that an unstabilized free fall from above 100,000 feet was achievable.

As Nick had confided to Senator Williams with supreme confidence many months before, the easiest part of the whole enterprise would be the jump itself. With a balloon contractor signed up and successful pressure-suit jump tests behind him, he could now enjoy the luxury of imagining that instant when he would step out into space and, for a few exotic, exhilarating minutes, inhabit the solo universe of his dream.

3 STRATO-JUMP I

Frustrations End

Nick remembered the name of Airman First Class Marvin McCall. The entire physiological training team at Tyndall had done a great job of preparing him for his sessions in the decompression chamber, but Nick had been especially heartened by the extent of this particular young man's enthusiasm for the mission of Project Strato-Jump. McCall, an affable, round-shouldered technician whose job involved training F-106 pilots in the operation of full-pressure suits at extreme altitudes, was a kindred spirit who seemed instinctively to understand Nick's personal motives in wanting to break the world free-fall record. He clearly believed in the larger goals of the project: to prove once and for all the survivability of stratospheric skydiving without the use of a stabilization device like a drogue chute and to make an international statement about the capabilities and continuing ambitions of the United States in super-high-altitude endeavors.

Following his visit to the FAA in Oklahoma, Nick contacted McCall with an update on his efforts to secure the David Clark suit, saying that he had decided "to continue the project with as little Air Force participation as possible." At the same time, Nick made it clear that while he intended to distance himself from the Air Force, he very much wanted McCall involved in the project.

On July 2, as he began the process of staffing the ground crew for Project Strato-Jump, hand-picking individuals with the competencies required, Nick presented McCall with a written proposition: "I would like to have you as the pressure suit technician on my ground support crew. The only way I think this can be arranged is if you could take a leave. I will pay the round trip air fare for you and your wife, and the motel lodging plus meals. In addition, I will *give, not pay,* you $300 for your services." This arrangement, presented with a full measure of the hard-sell Piantanida charm, was intended to compensate McCall without creating the appearance of official Air Force involvement. Nick outlined his plans for a September launch from Minneapolis and said that he would need McCall on site at least three days prior to launch. He also asked McCall to keep quiet about the $300 arrangement. The rest of the crew would be working as volunteers.

McCall got in touch with Nick almost immediately to accept the offer, saying that he had already scheduled the leave with his superiors who remained privately supportive of Nick's efforts. "I'm sorry the Air Force gave you a hard time about the suit," McCall wrote in a letter to Nick, "but they have a way of doing that to everyone." He also mentioned the financial arrangement Nick had proposed, making it clear that he considered the $300 "gift" to be quite generous: "As you know, we service guys don't get deals like this very often."

Courting NASA

In addition to overtures to the Air Force, the Navy, the NAA, and the FAA, Nick had attempted to court one other federal organization that he believed might have an interest in his dream: the National Aeronautics and Space Administration. Nick had first contacted NASA in November of 1964 in his bid to obtain a pressure suit and had been informed that NASA obtained all of the suits for its astronauts through the Air Force. Now that he had a suit at his disposal, as well as a contractor to build him both a balloon and a gondola and to launch him, he made a different kind of pitch to the nation's space agency.

At the suggestion of George Johnson at the David Clark Company, Nick approached Dr. Charles Berry who was in charge of medical programs at the

Manned Spacecraft Center in Houston and asked for any technical support NASA would be able and willing to provide. In return, Nick offered to modify Strato-Jump's agenda to include any biomedical experiments or other data-gathering that might be useful to NASA, generously promising to forward all project results to the organization regardless of whether it was willing to cooperate openly. It wasn't farfetched to imagine that NASA might be interested in taking advantage of opportunities with Strato-Jump; the agency was already actively involved with a number of private contractors and with both private and public universities (including the University of Minnesota) researching the severe conditions in the near-vacuum of space. The essence of the NASA response, however, was that while the agency found the project's objectives interesting and of great potential value to the United States, they were "not in consonance with present requirements of NASA."

Not much later, Jacques Istel made his own overtures to Hugh Dryden, NASA's deputy administrator, again seeking the agency's cooperation and not solely in the area of funding. Istel presented Strato-Jump as a research project with a potential airborne laboratory for the study of high-altitude survival and proposed teaming up with NASA personnel for live tests of rocket escape systems. Dryden was clearly intrigued and offered to pass the offer on to George Mueller, the associate administrator for manned space flight, and to Bob Gilruth, director of the Manned Spacecraft Center.

Shortly after Istel's conversations with Dryden, Nick was granted an appointment with NASA officials in Washington—an arrangement that may have owed as much to the efforts of Senator Williams as to any actions taken by Dryden. Mike Quigley, a recently hired management intern at NASA, recalls escorting Nick to the agency's congressional liaison office. Quigley remembers Nick as clean-cut, quick-witted, enthusiastic, and generally likable, and recalls him talking nonstop about the research potential of manned high-altitude balloon flight. Nick had, of course, done his homework and was by this time a well-practiced evangelist ready at the drop of a hat to make the case that, by ignoring Strato-Jump, NASA would be turning its back on a golden opportunity to gather valuable data for a minuscule investment (essentially the same argument for balloon-borne research platforms advanced so eloquently in earlier years by Otto Winzen and Malcolm Ross). And Nick cut a formidable figure on Capitol Hill, coming across as something like a well-spoken professional athlete turned lobbyist. Dressed in his dark blue suit that did little to hide his massive build, and armed with the ever-present briefcase, he strode across the wide boulevards of the capital in what sometimes appeared to be three-quarters speed. In spite of his tremendous reservoir of energy and lizard-quick reflexes, Nick never seemed to be in a hurry no matter how urgent the mission.

Following Nick's appeal to NASA, which was ultimately unsuccessful, Mike Quigley was waiting to escort him back out of the building. It was Quigley's impression that not only was Nick completely undaunted by the lukewarm reception and resolute in the face of yet another potential disappointment, but that his confidence in himself and his project emerged even stronger than it had been going into the meeting.

Nick, his obsession more maniacal than ever, now full of optimism and gaining confidence in his project, was beginning to believe that he no longer really needed the support of behemoth organizations like the Air Force and NASA. If they didn't want an association with Strato-Jump—well, it would be their loss. Nick believed that he was finally in a position to define success on his own terms.

Throughout the late summer and early fall of 1965 Nick worked the angles—and there were dozens of them—relentlessly. He chatted up every reporter who would listen in an attempt to generate publicity and to attract additional financial contributions. He continued his consultation with Firewel engineers on the various components of the oxygen system, which the company had now formally agreed to loan to the project. Still to be determined was whether Strato-Jump would employ an all-gas system or if it would go with liquid oxygen and a LOX converter. Pioneer Parachutes was completing modifications to the specialized Para-Commander Mark I canopy, drogue chute, and special harness rig. At the same time, Nick had to continue applying for thirty-day leaves of absence from his trucking job with Smith Transport in New Jersey in order to retain his standing with them. Regardless of what was happening with the project, there was always rent to be paid and a family in Brick Town who needed to eat.

Modifications, mostly thermal-protection enhancements, to the Dave Clark Special had continued apace. Based on the results of the cold-chamber testing in Oklahoma City, the pressure-suit engineering team in Worcester decided to add a fifth layer to the suit's coverall fabric, this one of neoprene. Jacques Istel had phoned Joe Ruseckas, David Clark's vice-president of research and development, in late August. He said that he, along with Pioneer Parachutes, was now officially involved with Project Strato-Jump and that there were concerns about pressure-suit performance at stratospheric terminal velocity. Specifically, Istel inquired about the company's confidence in the ability of its suit to withstand the forces of Mach 1.1 that Nick expected to encounter on his free fall. While Ruseckas was unwilling to offer Istel any guarantee, he did forward the results of Air Force rocket-sled experiments conducted four years earlier on an A/P22S-2 suit that was subjected to supersonic wind blasts of Mach 1.08 with satisfactory results. A further test had been conducted using a high-speed dummy ejection from an F-106 that exposed the suit to a wind blast of Mach 1.58 (approximately 730 knots) without any measurable degradation in performance. In a memo to John Flagg, Ruseckas speculated—accurately, as it turned out—that Istel's ultimate motive was to "sell a development contract to the government."

As a result of the modifications to the suit, Nick felt that he needed a final decompression chamber test with the full survival system. The FAA, the most cooperative of all the federal agencies to this point, quickly scheduled Nick for another session in late September. An official in Oklahoma City who had become familiar with Strato-Jump and with Nick commented that he was confident the project would "contribute new knowledge to the aerospace sciences." The September date was not a schedule problem for Strato-Jump because delays at Litton had already forced a postponement of the launch into October.

On September 8 Nick took the train into Manhattan and signed a contract with SPACE, Inc.—essentially an agreement between Nick and Jacques Istel—that anticipated not only a successful project but its potentially lucrative aftermath. The contract bound Nick's activities to the corporation and prohibited him from profiting individually from any opportunities created by the proposed world-record jump. The agreement stipulated how income generated from television and personal appearances, lectures, and patents would be apportioned. Proceeds from the jump, after expenses had been covered, would go 80 percent to Nick and 20 percent to the company. Should there be subsequent jumps, 30 percent of the profits would go to Nick and 70 percent to the company. Istel wanted to ensure that any windfall created as a direct result of SPACE, Inc.'s investment would be used not only to reward Nick but to sustain and build the venture for the future.

Istel and Piantanida had often discussed the future of SPACE, Inc., and both men were convinced that there was in fact an excellent potential for continued and lucrative high-altitude research, perhaps from military contracts (as Joe Ruseckas had guessed), beyond the world-record jump. For example, their contract stated that Nick would reserve the right to make the first jump from "any rocket operated by the company," a reference to a one-man rocket-powered vehicle that Nick had already designed in his head and which he imagined might be a suitable follow-up to Strato-Jump. According to Janice, "Nick always said he wanted to be the first man on the moon."

But SPACE, Inc. would be more than a launchpad for Nick Piantanida's fantastic dreams. As an inventor, Istel had ideas of his own and saw the company as a way to contribute to the design of new emergency escape systems. He envisioned a system of huge parachutes that could be used to bring disabled airliners safely to earth: "In a mid-air collision, for example, where a wing was sheared off, the passenger cabin might be separated into sections and lowered by chutes."

Another important issue addressed by the new contract was liability. SPACE, Inc. agreed to insure Nick's life for $140,000 ($100,000 payable to Nick's estate, $40,000 to the corporation), and Nick agreed to carry his own medical and accident insurance, and to indemnify the company against any claims. An additional clause required that Janice agree to indemnify the company as well.

Two days later, SPACE, Inc. signed the final contract with Litton Industries for the Strato-Jump balloon and launch services. And on October 15 SPACE, Inc. would sign a deal with Bob Perilla and Associates in New York granting the firm exclusive public relations rights to the project, excluding TV rights to the jump and a postjump first-person story for *True* magazine, which had already been negotiated.

Nick's second visit to the Aeromedical Center at the FAA facilities in Oklahoma City on September 24 went exceptionally well. Using a new Firewel regulator and the Dave Clark Special, now with beefed-up thermal protection, he made a chamber run of one hour and twenty-five minutes to a pressure equivalent of 127,000 feet. In attendance were flight support technician John Fitch, physiologist Dr. Stanley Mohler, with Joshua Mann serving as chamber operator—the same crew Nick had worked with in June. But this time Nick was accompanied by Barry Mahon and a small crew who filmed the entire proceedings through a glass portal on the big FAA tank.

Following a twenty-minute, 100-percent-oxygen prebreathing session, Mann dropped the pressure in the chamber, slowing briefly as the pressure equivalent approached 40,000 feet to verify that the Dave Clark Special had fully inflated and was holding pressure. Nick reported being quite comfortable, with the exception of some excess heat coming from the bright lights Mahon had set up inside the chamber. Nick experimented with his ventilation unit, even though he found that he hardly needed it.

At each thousand-foot mark, as standard chamber protocol, Mann would read out the simulated altitude and ask Nick to respond with the reading from his suit altimeter, which was holding steady at between 33,000 and 35,000 feet as designed. Nick complied for a while, but as the ascent reached 45,000 feet, he stopped offering the suit reading. When Mann requested the reading, Nick told him just to assume the suit pressure was in the acceptable range unless he told them otherwise. Mann explained that he needed Nick's responses for the tape-recorded record, but Nick countered that it was "too repetitious," and Mann graciously let the matter drop.

About ten minutes later, Mann asked Nick how he was doing.

"It's a little boring," Nick replied. Even though the chamber run nicely simulated the period of inactivity Nick would experience as the pilot of a high-altitude balloon, the lack of stimuli taxed his patience.

Mann explained, "That's why we like to talk to you."

At 120,000 feet, Nick practiced switching from his main oxygen supply to a bailout bottle in his backpack to simulate preparing for the jump. As he performed this task, his respiration began to speed up and Mann asked him to slow his breathing. It was a request Nick would hear numerous times later on his actual trips into the stratosphere.

While at the pressure equivalent of 127,000 feet, Mann asked Nick to experiment with a series of body movements to test the mobility of the Dave Clark Special under the restrictions of full pressurization. Nick flexed and swung his arms around for a while, pronouncing the result satisfactory. At this point, Barry Mahon took Mann's microphone and addressed Nick inside the chamber: "You wouldn't be able to move around at this altitude with this much pressure in the Arrowhead suit, you know," he said.

Mann next gave Nick a quick "descent" from 120,000 to 50,000 feet to simulate free fall. Nick stood up for part of this section of the run, prompting Mann's most alarmed comment of the session: "Don't get too close to that damn light. We don't need a fire with all that oxygen in there."

On the final leg of the simulated descent, Mann informed Nick at 27,000 feet that he could safely open the visor on his helmet at any time, but Nick responded that because it would be so cold at that height, he wanted to practice leaving the visor down until shortly before landing.

At 10,000 feet, Nick conceived of an antidote to boredom and asked to have the chamber pumped quickly back up to a 50,000-foot equivalent.

"Hey, Barry," he called over his mike, suddenly very engaged, "get a film of this. Rapid decompression!" Mann took Nick up to the equivalent of eight miles in forty seconds. The suit pressurized perfectly and Nick remained comfortable inside it.

"So, are you going to buy the suit now?" Mann joked.

"Yeah," Nick said, chuckling, "I'll take it."

Following this final test of Strato-Jump's survival system, the FAA's Dr. Mohler issued a plea—at Nick's request—to an Office of Naval Research representative at the University of Minnesota, encouraging ONR to consider providing support vehicles for the launch planned for the following month. Writing on behalf of the Civil Aeromedical Research Institute, Mohler assured ONR that Nick's project was well thought out and was to be taken seriously: "We feel that this project will produce information of major significance to air safety. . . . Any cooperation extended by your Department, especially with respect to la unch vehicles, will be very much appreciated, and will be of significant assistance to the accomplishment of this important project." ONR subsequently agreed to loan not only a launch truck but several ground vehicles to the project, vehicles that might otherwise have had to be rented or bro ught in from other parts of the country. The Strato-Jump budget would depe nd on these sorts of favors.

On another front, Nick was personally negotiating with the Capp Towers Motor Hotel in Minneapolis for lodging for the crew and for facilities to handle press conferences both before and after the flight, and for a private meeting room large enough to handle the team's victory celebration. In return for discounted rates, Nick promised to mention the Capp Towers prominently in press releases and to include exterior shots of the building in the feature film Barry Mahon would be making. He had similar conversations with rental car companies, commercial airlines, restaurants, and a whole roster of equipment vendors.

"The Great Balloon Race"

The big jump was now scheduled for the third week of October. Nick and Janice had traveled to Minneapolis earlier in the month for Nick's formal balloon pilot training. Before he could legally lift off as a solo pilot in any type of balloon, he would need to earn a free-balloon pilot's certificate. He had discussed both hot-air and gas-balloon training with Don Piccard that summer and had originally agreed to a ten-day session with Piccard in Fort Dodge, Iowa, in July, but Piccard's consulting fee of $100 a day had caused Nick to reconsider. Deciding finally that a gas balloon would simply be too expensive for training, Nick had contacted one of America's top sport balloonists, Tracy Barnes of Wayzata, Minnesota, to teach him to fly hot-air balloons. Barnes, a fellow free spirit and former physics student at the University of Minnesota who would later found his own balloon manufacturing company called Balloon Works and who had made a flight to 38,650 feet in a sport balloon of his own design the year before (a flight that captured thirteen officially sanctioned altitude records), was a good choice. He prepared a brief ground-school session for Nick and supervised the required six hours of instructor-led flight training. The training with Barnes was both fun and relaxing for Nick, and provided him a brief respite before the whirlwind that was just ahead.

The final requirement for the balloon pilot's certificate, however, presented something of a problem. Nick needed a minimum of one additional hour's flying time without an instructor in the gondola, and in spite of Nick's powers of persuasion Barnes was unwilling to surrender his prized 43,000-cubic-foot balloon to a novice. Too much could go wrong. "I was fairly impressed with Nick's drive and sheer determination," Barnes would recall. "He was a confident guy. And he was quite competent with the balloon." On the other hand, Barnes also sensed that Nick was just a little *too* gung-ho and refused his student a solo flight.

So on a beautiful fall morning, Nick arranged one final purely recreational flight with his instructor. "It would be great if we could take Janice up in the balloon so she can see what it's like," he said. After a leisurely flight, Barnes put his balloon down in a brush-covered field to change propane tanks. And then somehow, with the balloon still inflated and both Nick and Janice holding steady in the gondola—and Barnes on the ground attending to last-minute chores—the whole thing got loose and began to rise.

Barnes screamed in horror, "Bring my balloon down now!"

Nick smiled and shrugged in mock helplessness.

"Nick, we're not supposed to be doing this," Janice said.

Nick replied, "But I have to get the hour in. Honey, all we have to do is stay up for one hour. Just one hour."

Barnes and his chase crew took off after the balloon in a pickup truck with Barnes yelling out the window at Nick the whole way—and Nick hollering back, even as he fired the burner to keep the balloon aloft and skimming the tree tops, "Just one hour! That's all I need, Tracy! I'll put it down in one hour, I promise!"

Before long, a Minnesota Highway Patrol car with red light flashing had fallen in behind Barnes's truck, which continued to chase the balloon across the gently rolling plains of the Minnesota countryside.

"I'm really sorry!" Nick continued to yell down to the ground convoy. "I'm trying to bring it down!"

Janice was worried at the time that Barnes might try to have Nick arrested for stealing his balloon. "I was pretty peeved," Barnes admits. Later, Janice could laugh. "We called it 'The Great Balloon Race.' Of course, Nick knew *exactly* what he was doing. He landed the balloon without any damage exactly an hour later. It was a wonderful flight. I loved it." And while Nick may have burned his bridges with Tracy Barnes, he simply felt he had no other option. Barnes, either out of graciousness or simply to rid himself of Nick for good, agreed to sign off on the solo flight and placed his signature in Nick's balloon logbook. Now Nick had his certificate. He was primed, ready, and street-legal for Strato-Jump.

During the early part of their time in the Twin Cities, Nick and Janice accepted an invitation to stay at the Minneapolis home of Litton Industries' Karl Stefan and his wife, Lucy. It was a serendipitous arrangement. Both of the Stefans were accomplished balloonists and engineers with deep roots in American aeronautics. Karl was a former jet pilot and a veteran of General

Mills' scientific balloon research and Lucy had flown coal-gas balloons in Philadelphia with Don Piccard, had studied under Piccard's father Jean, the prototype absent-minded professor, at the University of Minnesota, and later flew hot-air balloons with Tracy Barnes. The Stefans were both completely enchanted by the Piantanidas.

"He was a delightful, charming gentleman," Lucy says of Nick. "Just so nice to everybody. My kids adored him. Every night he'd play on the living room rug with them. He'd throw them in the air and wrestle with them for hours, never tiring. And Janice was the gracious wife. You could see that the two of them were devoted to each other."

Once the children were in bed, the Stefans and Piantanidas would swap stories into the night. Lucy was particularly interested in Nick's accounts of Angel Falls and his Venezuelan adventures and made her own lasting impression on him by revealing that while working for a South American airline in the 1940s she had gotten to know an audacious bush pilot named Jimmy Angel.

Most of the rest of the crew began arriving in Minneapolis on the weekend of October 16–17. The launch date had originally been selected so as not to interfere with or overlap the publicity arrangements for NASA's *Gemini 6* flight, but as the Gemini launch had itself been scrubbed and moved out—it was now scheduled for October 26—Strato-Jump felt confident in targeting Friday the twenty-second for its own liftoff.

But there was another potential complication. The weather in Minnesota was not cooperating. Clouds, showers, and cool temperatures prevailed on Saturday and Sunday, with more clouds rolling into the Twin Cities area on Monday and Tuesday. On Wednesday it rained steadily with daytime highs in the seventies. Then, on Thursday, a high-pressure system finally swept in from the west, bringing clear skies and high spirits to the balloon launch team. With the uncertain weather, some thought had been given to moving the launch site seventy-five miles to the east to Eau Claire, Wisconsin, but plans were finalized on Thursday morning to launch as originally planned from the old University of Minnesota Airport off Highway 8 in the town of New Brighton, just five miles northwest of St. Paul.

Everyone was in town now, and the Piantanidas had moved into the project's unofficial headquarters. Staying at the Capp Towers Hotel with Nick and Janice were Jacques Istel, Barry Mahon and his wife, Clelle (in a suite), Marvin

McCall, Charles Alexander (the chute rigger supplied by Pioneer Parachutes), John Fitch and Joshua Mann (FAA), and Phil Chiocchio, Dan Quinn, and Bill Hennigar (all Lakewood jumpers who had signed on as part of Mahon's film crew). In addition to Litton personnel and the group at the Capp Towers, a number of other members of Nick's extended team were scattered around town. Bob Perilla and George Fenmore, from the newly hired PR firm, had arrived. Don Piccard was serving as an observer for the NAA along with two other officials from the Parachute Club of America. Ed Vickery was representing Pioneer Parachutes' management. Maj. Chuck Dorffeld of the U.S. Marine Corps was coordinating helicopter operations. The Marines, at ex-Marine and Korean War veteran Jacques Istel's request, had generously offered not one but two helicopters (complete with crews) for tracking the balloon during the ascent and for tracking Nick during his fall. With the addition of a civilian helicopter, the Navy ground vehicles, the Litton chase truck, and Barry Mahon's film van, Strato-Jump had transformed itself into a veritable guerrilla army poised for an assault on the stratosphere. In total, including both the Strato-Jump and Litton crews, some fifty individuals would be actively participating.

The Strato-Jump crew wasn't in the Twin Cities long enough to become familiar with all of the nightlife, but they found their share of relaxation in the evenings. One night Barry Mahon—who seemed to have interesting and useful connections everywhere—treated a large contingent to dinner in the rotating restaurant atop the Capp Towers. After dinner the group moved on to a local honky-tonk to unwind and revel in the western harmonies of Pat Brady and the Sons of the Pioneers. Between sets Mahon escorted Brady—who in the 1950s had been well known as the cowboy star's sidekick on the *Roy Rogers* television show—over to the table and introduced him to Nick and the others. It was one of those nights of laughter, liquor, and camaraderie that a team like Project Strato-Jump needed every so often to break the tension and to preserve its collective sanity.

Janice recalls Nick introducing her to Don Piccard ("an odd fellow") and to Don's remarkable mother, Jeanette ("a really, really lovely woman"). Nick and Janice visited the venerable Piccard family house on River Road which provided a symbolic link to the glory days of gas ballooning. Mrs. Piccard apparently warmed up to Nick right away, as if he reminded her of the adventurers of earlier times.

"I think she helped Nick a lot," Janice says. "She really took a liking to him and I think she was beneficial in a number of ways. I think maybe she felt like Nick was a long, lost son. But all the people there, especially the Litton people, were so good to us. And they would all keep me busy during the day, arranging for lunch at different locations, taking me around and showing me the sights. Maybe Nick asked them to do that. I don't know."

Nick was clearly sensitive about Janice's apprehensions, but he had insisted on a full-disclosure policy in terms of information about the project and the flight. He made a solemn promise to Janice that nothing was to be withheld from her, even if the flight went badly. He was adamant that she should have access to the communications van during the flight with one stipulation: she was not allowed to try to talk to him over the radio once the balloon was airborne. Everything at that point would be strictly business.

"I promised," she says, "but the truth is I did talk to him once or twice."

Once the balloon was launched, flight control would be provided to three individuals: the ground-control monitor, the lead tracking pilot, and the balloon pilot himself. In most circumstances, ground control would be running the show. But a key protocol decision was reached at one of the final preflight briefings. In the event of a total communication breakdown—and this would hardly be unprecedented, as radio problems had beset several of the great Air Force stratospheric balloon flights—the pilot would assume full command and would accept the responsibility of dealing with any emergencies completely on his own. In the absence of communication with the gondola, ground control would only terminate the flight after the balloon had remained at float altitude for one hour.

Nick had put the finishing touches on the master flight plan on October 12 and, with the help of Litton personnel, had completed the preflight checklists two days later. Both sets of documents had been copied and distributed to all members of the team. Each function had its own formal list of tasks and schedules.

There was some eleventh-hour red tape to cut. On October 14 the local FAA office in Minneapolis issued a registration number (N7534U) for the balloon and a set of operating restrictions governing the impending flight. Three days later Nick applied to the Minnesota Department of Aeronautics for approval to make a high-altitude jump in Minnesota airspace and received written approval the following day. That accomplished, the bureaucracies having been fed, the only potential roadblocks in Nick's way at this point were bad weather, bad judgment, and equipment failure.

Nick had been giving regular interviews to the Twin Cities press since his arrival earlier in the month. On October 16, after renting a Hertz van that his support team needed, he posed for photographs in the David Clark suit, seated in the gondola. His press conference in a banquet room at the Capp Towers on Wednesday was, as the local reporters had come to expect, a great show of bravado. "I should reach the speed of sound—Mach 1—at about 100,000 feet," he told them, "and Mach 1.1 at about 89,000 feet." Nick also assured the press that he had "government support" for his project, although he declined to elaborate.

That same day Bob Perilla spoke on the telephone with a reporter from the *Asbury Park Evening News* back in New Jersey and admitted, a bit grudgingly, that Nick had at least considered some of the problems that had plagued earlier high-altitude parachutists. "We discussed other accidents that different jumpers have had," Perilla said, "but Mr. Piantanida was unconcerned." While it's only a bit of a stretch to believe that Nick himself was totally happy-go-lucky in the days leading up to the flight of Strato-Jump I, it's impossible to credit Perilla's further comments that Janice was "similarly unworried." Part of the project's public relations strategy was to portray the couple as unfazed and carefree, a strategy that would become increasingly untenable in the days, weeks, and months to follow.

All that week both Karl Stefan's Litton crew and Nick's Strato-Jump team worked through their checklists and gathered for a seemingly endless series of briefings and conferences. Oxygen, in both liquid and gas form, was procured and delivered to the Litton facility. (In consultation with Firewel, Nick had elected to use liquid oxygen for the suit ventilation system.) Both the onboard breathing-oxygen tanks—as well as the two smaller 60-cubic-inch bailout cylinders Nick would carry with him as he leapt from the gondola—were filled and topped off. The onboard oxygen supply consisted of an 875-cubic-inch main cylinder and a 295-cubic-inch reserve cylinder, both pressurized to 1,800 psi. Together, they would provide a total of nineteen hours of breathable oxygen in the gondola, significantly more than would be needed under almost any flight scenario.

The ventilation system, a portable cryogenic unit with three liters of liquid oxygen sufficient for five hours of suit cooling, was checked. Instrumentation for the gondola, bailout system, and ground vehicles was calibrated, checked, and double-checked. Aircraft and ground communications and telemetry systems, including a voice-command transceiver, a 225-megahertz open-microphone transmitter, a barotransmitter, and a command receiver for dropping both ballast and the antenna, were tested. The tape recorders, one at monitor control to record in-flight conversation and another that would be secured to the pressure suit to record Nick's commentary during free fall, were inspected, cleaned, and tested.

The electrical system, two 29-volt silver-cell batteries for onboard instrumentation and radio power and a nickel-cadmium battery with 30-volt and 15-volt taps for power during the jump (tucked into a pocket behind the reserve chute), was tested. Other jump gear was carefully inspected: the jump altimeter and a stopwatch (in a panel on top of the reserve) and a second stopwatch that would be worn on the left wrist. The entire system was weighed and photographed. Litton security personnel were furnished with rosters of individuals who were to be admitted to the working area immediately surrounding the launch site.

For Nick, it was marvelous to see the operation running on its own, without his constant intervention. Perhaps he was able to appreciate the irony of the supreme loner, the solo artist, as the architect and leader of his own team. The whole collective motor was humming now, the project was cruising along, and for brief moments he was able to take his hands off the wheel, sit back, and simply observe. Not that his constitution would tolerate such a luxury for long. Most of the time, night or day, he could be seen hustling about the plant looking for problems to solve or double-checking the work of his team. He believed that it was important, as the project leader, to be a physical presence in the Litton plant.

On Thursday evening Nick and Janice listened to an optimistic weather briefing and turned in early in their fourth-floor room at the Capp Towers, anticipating a 3:30 A.M. wake-up call. While they slept, Tim O'Malley, the project engineer who would monitor the flight from the Litton telemetry bus and who—along with balloon operations manager Karl Stefan—would have access to all flight controls, had a final conversation with meteorologists at the FAA. The FAA's clearance for the flight of Strato-Jump I was contingent on certain weather minimums: less than five knots ground wind, less than forty knots upper-air winds, and no more than 5 to 10 percent cloud cover. And now, with a forecast in hand that satisfied those conditions, FAA officials gave Litton the GO to launch later that morning.

O'Malley was a technician rather than a degreed engineer, but he was an extremely capable individual with plenty of hands-on operations experience, and Stefan trusted him to handle the day-to-day business of the balloon group. The GO order from the FAA triggered phone calls from O'Malley to the members of the launch crew, ordering them to the Litton plant in St. Paul. They arrived an hour later and departed for the airport thirty minutes after that.

Point of No Return

At 6:00 on the morning of October 22, one hour prior to the scheduled launch time, inflation of the big balloon began. This was a routine operation familiar to the Litton crew, who maneuvered the launch truck containing the gondola into place at one end of one of the runways and an anchor vehicle at the other and began rolling out the nearly 120,000 square feet of polyethylene along the canvas carpet laid down to protect the delicate plastic. They connected a hose from the helium trailer to a polyethylene duct on the balloon called an inflation tube and began releasing the pressurized gas into the top end of the envelope, the ghostly whine announcing to the approximately one hundred spectators and nonessential crew lined up along the edge of the runway that the point of no return loomed just ahead.

As Tim O'Malley coordinated the various teams of technicians, Karl Stefan found himself frustrated by a phenomenon unfamiliar to him: a ravenous press corps in free-for-all mode with little regard for protocol and which took orders from no one.

"One of the things I didn't anticipate was how aggressive the news people would be," Stefan admits. "They were terrible. They were just running around. I was afraid they'd step on the balloon. They'd run in and out of the van while Nick was attempting to suit up. And I realized I couldn't stop them. They were just uncontrollable." Stefan resolved that on future high-profile flights he would devise a way to keep the press fenced off from the launch site. But it was too late for that now. All he could do was try to keep them from trampling the precious polyethylene and hope for the best.

There was a long list of last-minute chores and equipment checks. Helium calculations were double-checked: 25,937 cubic feet of helium would be introduced into the balloon, which would expand to 3,705,000 cubic feet of gas at float altitude. A technician was assigned to make sure that the balloon's load tapes—which would bear the weight of the payload—were evenly distributed as a relatively small portion of the envelope was fed through a big roller called a launch spool prior to being laid out, while another man inspected the polyethylene and watched for holes, defects, or stress marks in the material. All release squibs (small explosive charges) were positioned and armed.

The cargo chute that physically connected the balloon to the gondola was checked to make sure that all attachments were sound and that extraneous tape had been removed from the chute's shroud lines. Both of the instrument panels in the gondola were checked to verify that all plugs were firmly connected. The left-hand panel contained all the electrical power switches for control of ballast (the flight would carry two forty-pound containers of liquid ballast), antenna, helmet visor and undergarment heat, camera activation, and balloon release. The right-hand panel held temperature gauges and an altimeter.

One of the final tasks was to connect a shock pad to the bottom of the gondola. Even though there were no plans for the pilot to land inside the gondola, Litton wanted to cushion the fall to protect the instrumentation and the integrity of the gondola itself in case the project would need to use it again. Just moments before the Litton team withdrew from the gondola, two last-minute emergency items were stowed aboard: a flashlight and a pair of wire cutters.

With just thirty minutes to go, the gas bubble in the balloon envelope began to bulge and rear up above the anchor vehicle like a living thing, as amorphously graceful as a jellyfish in a swirling current. Simultaneously, a Cessna 170 with Litton pilot John Benson aboard took off and began a slow circling of the airfield. Benson's was the primary tracking aircraft and his job was to circle the balloon's coordinates during the entire ascent, although he

The Litton balloon is inflated at dawn. (Courtesy *St. Paul Pioneer Press*)

would be far below for most of that time. Even after Nick's jump, Benson would continue to circle the gondola's position and track it all the way back to the ground. A second airplane, a Cessna 180 flown by Clarence Hines, was designated the parachute recovery plane. Two Lakewood jumpers—Dan Quinn and Chuck Alexander—would be aboard that plane and would be ready to parachute to Nick's assistance in the event of an emergency. In addition, the two helicopters provided by the Marine Corps were readied for recovery of the pilot. The Marine crews, led by Helicopter Squadron Commander Major Dorffeld, were the first option. The parachutists and the ground vehicles would be utilized only if Nick's landing site was inaccessible to the choppers. The Marines were also equipped and prepared to administer first aid in the event of an injury to Nick. A local hospital with a hyperbaric pressure chamber was on standby in case Nick experienced a failure of his survival system and suffered decompression.

As the Cessnas began their climbs, Nick—fully suited in the Dave Clark Special and supplemental garments—kissed Janice, fended off the more aggressive reporters, and climbed into the gondola which bore a new designation on the placard pasted on one of the exterior side panels. In the bright spot of light provided by Mahon's film crew, it read: FRUSTRATIONS END.

Nick took his place on the plywood and foam-rubber seat while technicians hooked up the oxygen and electrical systems in the gondola to connections on the pressure suit. At approximately twenty minutes prior to launch—it was now 7:00 A.M. and a weak sun was spreading muted banners of pink and gold across the low eastern sky—Nick sealed the pressure visor on his helmet and began to breathe pure oxygen from the main tank in order to purge his bloodstream of nitrogen.

At ten minutes to launch, Tim O'Malley huddled one last time with his launch crew before announcing the final GO decision to launch. The barographs were activated and O'Malley shared a few words of encouragement with Nick over the radio. At five minutes to launch, the ground-recovery vehicles fanned out toward the predicted landing area in western Wisconsin. With three minutes left Nick gave one last thumb's-up, and the gondola door was lowered. Nick was now completely enclosed in the white box and, although he could see out through the small windows on three sides, he was essentially invisible to all but the closest observers. Janice, with what would become her trademark scarf covering the dark hair piled on her head and obscuring as much of her face as she could manage, backed bravely away, her eyes welling with tears. She had been imagining this moment for well over a year now, but she nevertheless found herself nearly overwhelmed.

At precisely 7:20 A.M. Central Standard Time, the launch spool pivoted to release the balloon bubble which rose and lifted the uninflated remainder of the envelope. The launch truck then began rolling in an attempt to match the direction and speed of the gentle breeze. The truck had to veer off the runway and head off across the rough infield, which created a bumpy few moments for Nick. Once the elongated envelope was fully vertical above the launch truck, squibs releasing the whole assembly were fired and the great balloon was free. It rose initially at a rate of approximately 1,200 feet per minute, several hundred feet per minute faster than they had intended. Karl Stefan, a former Navy commander with extensive experience in aerospace research and scientific balloon operations, monitored the ascent and, by controlling the ballasting operation, managed to slow the rate of rise down to 1,000 feet per minute. At that rate, the ascent to float altitude, somewhere above 115,000 feet, would take less than two hours.

Once Nick had completed his initial communications and oxygen checks, with all systems behaving as expected, *Frustrations End* made its first voice transmission to the ground. "Sure is quiet up here," Nick reported. Tim O'Malley, a former Navy man whose specialty was flight instrumentation and communications, was now in charge in the communications van and it was his voice that answered.

"Roger. Nice launch, Nick."

"Sure was a beauty, Tim," Nick replied. "A little rough when you ran over that grass though." (The Litton launch team had used ONR's new horizontal dynamic launch vehicle. It was the first time it had been used for a manned flight.)

Jacques Istel was also in the van. "Everything looks good, Nick," he reported.

"Right-o, Jacques."

Janice, already breaking her pledge, was the next to join in: "God bless you, honey."

"Thanks, babe." Nick kept the front door to the gondola lowered but was able to maintain visual contact with the ground through the small plastic windows. "Nice view of Minneapolis. We should have hired out to traffic control this morning."

Now that the launch sequence was complete, Karl Stefan took over as the voice of Litton ground control.

"NCA-1," Nick called with a bit of concern in his voice, "this is NCA-4. Did you see anything come off the bottom?"

"The antenna just dropped," Stefan answered. Lowering an antenna was standard procedure. "Did you hear a squib pop? Did they forget to tell you about that?"

"Uh, yeah," Nick said, not willing to get too bent out of shape about anything this early in the flight. "That's okay."

"The squibs will also fire when we drop ballast."

"Okay," Nick said. "I just thought I might have hit one of 'em."

A couple of minutes later Nick came back on line. "Tell Jacques that this sure climbs faster than that Norseman [at Lakewood]."

Istel's voice answered, jokingly: "Yeah, we're going to put balloons in next season, Nick."

A minute later and Stefan requested an altimeter reading.

"8,100 feet," Nick said.

"8,100, Roger. You had it set to sea level at the launch?"

"Yes sir," Nick replied. He seemed to prefer a slightly more formal, more military communication style now that he was finally airborne.

An overcast, still several thousand feet above the gondola, was of only minor concern at this point. An observer in one of the tracking aircraft received a request from the FAA to have Nick report the altitude of the clouds once he reached them, and he passed that request to the communications van.

THE END OF THE BEGINNING

Phil Chiocchio was still a senior in high school when he headed off to Minneapolis with Project Strato-Jump. By day he was the bored audiovisual guy who wheeled projectors around the hallways and threaded up the educational films for the science and geography teachers, but his extracurricular life was something else again. At age fourteen he'd met Barry Mahon in New York City and had begged his way into a job sweeping floors at Mahon's studio with the goal of learning the film business. Phil was a fast study and, by resolving to accept any task without complaint, quickly became a valuable resource for Mahon Productions—although for obvious reasons he never really explained to his parents the *kind* of films he was cutting his teeth on. An unexpected benefit of his association with Barry Mahon was an introduction to the magical world of parachuting. Along with Dan Quinn, another Mahon employee, Phil found himself putting in time at the newly opened Lakewood Center in the summer of 1963 doing odd jobs. He made his first jump the week after his sixteenth birthday. Not long after, Nick had appeared at Lakewood, and not long after that, Phil found himself on the roster of Project Strato-Jump. "This was a dream life for me," he would say. "My friends were all doing homework and watching TV. And here I was, working on this guy's space program. Can you imagine?"

In Minneapolis as a member of Mahon's crew, working at Litton Industries, Chiocchio felt like a genuine big shot. And now, with *Frustrations End* in the air and his duties for the day mostly complete, young Phil stood watching the balloon's progress with the formidable Don Piccard who was attending the launch as an official observer, having failed to receive an invitation from Litton. At one point Piccard leveled his gaze at the young man and pointed to the sky, remarking somewhat ominously in his distinctive, booming voice: "Watch what happens when the balloon gets to those clouds."

"Why?" Chiocchio asked. He was quite aware of the Piccard reputation and, thrilled to be in his presence, wondered what the balloonist's eye could possibly be seeing. "What's going to happen?"

Piccard proceeded to explain that he believed the design of the balloon to be fundamentally flawed. He, of course, had favored a more durable Mylar envelope but, given the choice to use polyethylene, felt that too much of the load would be concentrated on too small a portion of the balloon. The design used 300-pound load tapes—the tapes actually attached the plastic to the rigging and assumed the weight of the payload—made of a synthetic called Fortisan. The calculations all showed that the tapes should have been perfectly adequate strength-wise, but Piccard doubted their ability to hold the balloon if it were distorted significantly by a wind shear and the weight was momentarily transferred and isolated to one or two of the tapes. And that's what Piccard saw: wind gusts at the level of the clouds might pull the balloon dramatically out of its natural shape.

"The balloon won't hold together," Piccard told Chiocchio almost dismissively. "Watch."

Chiocchio found the conversation very strange and wondered if Piccard knew what he was talking about. If he did, why hadn't he warned Nick and the Litton team? What Chiocchio couldn't have known was that Piccard had in fact already issued such a warning to a top Litton executive in Los Angeles about what he saw as a disaster in the making. Nor was there any way Chiocchio could have understood the depth of Piccard's pride and the sting he felt at being a mere observer—as opposed to an invited guest of honor—on a potentially record-setting balloon flight in his hometown.

At 11,000 feet, Stefan requested a complete reading of all onboard instruments.

"Outside temperature is 42 indicated," Nick began, "and inside is 54 indicated. From the conversion chart, this puts the outside at zero [Fahrenheit], and inside at about 15 above."

"Oxygen pressure?"

"Oxygen pressure is good, it's coming in at 70 psi. Pressure on the reserve system is at 74. Checks out A-OK."

"Are you needing any suit ventilation yet?" Stefan wanted to know.

Nick, still relatively comfortable, answered, "No, not yet, sir."

"Roger. Have you opened the door to take a look around?"

"No," Nick said, "but that sounds like a pretty good idea."

"Let us know how it looks. And don't fall out."

Nick laughed. "No, that would be a pretty damn expensive 60-second delay [parachute opening]."

At about 12,000 feet, Nick swung open the gondola door and saw the world before him. It was a markedly different sensation than looking down from an airplane.

"Aw, that sure is pretty," he said, his softened tone bordering on reverent. Not only was the view completely unobstructed, but the relative silence and the absolute lack of motion sensation at this altitude was almost surreal. He inspected the ground directly below him. "I'm right over a lake."

The pressure suit wouldn't inflate until about 35,000 feet, so his movements were still relatively unimpeded. At about 20,000 feet, Nick pulled the door closed again and activated the camera heaters as he felt the temperature begin to drop. He could see the thin overcast scudding by just above him from east to west, but its progress seemed quite slow and was still no more than a minor concern at this point. The ground crew seemed a bit more interested.

"There's a slight overcast around," Stefan informed him. "I guess you haven't passed through it yet."

"No, not yet."

"Let us know when you go through the overcast, Nick. The FAA would like to know."

"Roger," Nick responded.

Several minutes more, and Tim O'Malley's voice broke in. "NCA-4, this is NCA-1. Have you reached the height of the clouds yet? Over."

"NCA-1, this is NCA-4. No, I'm just starting to come up pretty close to 'em."

"Nick," Karl Stefan resumed, "you're on oxygen, I presume, since takeoff. Is that correct?"

"On oxygen?"

"Are you breathing oxygen now? Is your faceplate closed?"

"Yes sir," Nick said. "It's been closed since ten minutes before takeoff."

Then, as Nick approached 23,000 feet, just as he reached the edge of the overcast, he felt a distinct tremble. He described it later as a "quiver," as if something big had bumped the gondola. The vibration actually shook him and lifted him off his seat for an instant. The gondola was equipped with an overhead portal that allowed Nick to monitor the rigging and the balloon assembly above him. Constrained by the helmet, it wasn't physically possible for Nick to crane his head back far enough to look straight up, but he could use the overhead window by looking into the four-inch mirror attached to the right wrist of the pressure suit.

Like all high-altitude gas balloons, the rigging at the top of the Strato-Jump gondola was attached to a big cargo parachute. The parachute's rigging was then attached, in chain fashion, to the balloon itself by means of the load tapes. In the event of balloon failure, the parachute would inflate automatically. In such an emergency, it was critical to release the balloon before it collapsed on top of the parachute and the gondola. When Nick glanced at the mirror trained on the overhead window, what he saw was not the balloon as he'd expected, but an open parachute obstructing the view, indicating that he was falling rather than rising. Something had definitely gone wrong and they couldn't afford much delay.

Nick reached for the termination switch that would fire squibs to release the balloon, but he didn't push it yet. He checked his altimeter and found that he was in fact descending slightly. His assumption was that the balloon must have burst. But the rate of descent was so slow that he immediately began to second-guess that assumption.

"NCA-1, this is NCA-4," Nick said. "We have a malfunction. Should I terminate?" There was only the slightest hesitation before Nick heard Karl Stefan's voice in his earpiece.

"Yes," Stefan, immediately concerned about the envelope tumbling down on top of the parachute, responded forcefully. "Yes." A few seconds later, he queried the gondola. "Have you terminated?"

Nick pressed the switch and released the plastic envelope—770 pounds of plastic that had cost him $5,000, not to mention another $2,000 or so for the helium inside it—to the gusty winds at 22,700 feet. He was less than eighteen minutes into his dream flight.

"I've terminated the balloon." Nick's voice was calm, businesslike.

"Are you separated, Nick? From the balloon?"

"Yes, sir."

"You'll have to take care where you pick to land." Stefan said, distress evident in his voice for the first time. "Are you over the city?"

"I'll take a look," Nick answered, peering through one of the side windows. "Shit, I'm right over the city!"

It was an unspoken but foregone conclusion at this point that Nick would have to parachute out somewhere over St. Paul with little time to assess potential landing spots.

Stefan's voice came on again: "You'll have to pick an opening, Nick, and go toward it. Let us know when you are going out."

"Yes, I will," Nick said, still showing no sign of panic or anger. "I'm just riding with it. She seems to be coming down okay. The wind's blowing to the east."

Nick stayed with the gondola for more than three minutes as he tried to determine the extent of the damage. He found it hard to believe that the big balloon had burst. His first thought was that he had simply hit a severe downdraft. He wondered if they'd acted too hastily in releasing the balloon.

As he fell through 19,000 and then 18,000 feet, he reacted almost mechanically, as he'd been trained. He unbuckled his seat harness, disconnected his on-board oxygen supply, and once again pushed open the front door.

"Nick," Stefan said, "let us know when you're going to jump."

"I'll give you a warning when I'm gonna jump," Nick confirmed. "NCA-1, is it okay to try for that airport?" Nick could see the main St. Paul airport some distance away. Stefan radioed to one of the tracking helicopters to advise the airport that a parachutist might be on his way down at any moment.

"Hey, Karl? This is Piantanida. I'm gonna go out and try to make it to that airport."

"You're going for the airport?"

"Right." This was Nick's last transmission.

Seconds later, at 16,000 feet, he was poised on the lip of the aircraft's doorway, gripping the aluminum frame. He hit a switch that activated rapid sequence cameras and pushed off with his legs. He dove hard away from the gondola. Once airborne, Nick quickly adjusted his free-fall position in an attempt to track toward the airport, but the big reserve chute and other equipment on his chest prevented him from making much headway. He realized that reaching the St. Paul airport under these conditions would be impossible.

To see acres of buildings below can be intimidating, even for an experienced parachutist. At 10,500 feet, much higher than he would have liked, Nick pulled his D-ring and deployed the Para-Commander, hoping to be able to use the chute to maneuver his way away from the heart of the city and back to the vicinity of the launch site. But stiff winds made that impossible as well. He could see the tracking aircraft circling below him and could make out the recovery helicopters moving in from the periphery.

Nick was heading in a southeasterly direction, but at about 4,000 feet the wind shifted and picked up, blowing him due east. Soon there were only four choices for a landing: a housing development, a railroad yard crisscrossed with high-tension lines, the Mississippi River, or a nondescript stretch of land off Childs Road in St. Paul.

The houses and the railyard were obviously out of the question. "It was a little too late in the year to go swimming," Nick would explain. He touched down, executing a perfect stand-up landing, among the eggshells and rotting newspaper of the Pig's Eye Dump. A Marine recovery helicopter landed less than thirty seconds later. "Not very glamorous," Nick would say with a resigned shrug, "but that dump was the safest place to land."

It was hardly an auspicious beginning for what had become one of the most unlikely and audacious civilian aeronautical projects of the twentieth century. As Nick would characterize it in his parachute logbook: "Project Strato-Jump I: What a Bust!!!"

A man named Rex Wood had observed the flight of Strato-Jump from the ground using a high-powered telescope. He had seen no indication of a balloon failure before the big parachute strung between the balloon and the gondola began gradually to deploy. At some point between five and twenty seconds after the parachute was fully inflated, Wood noticed the balloon beginning to undulate, indicating a descent. A few seconds later, he saw the billowing envelope detach and lift away from the falling gondola.

Don Piccard's more experienced eye saw damage to the balloon occur before the parachute opened. It was almost exactly as he had predicted in the moments after liftoff: "It hit a bit of a shear and sort of distorted. And then it appeared to me that a banana peel sort of opened up and peeled away. I felt that, in this distortion, all of the load had gone on a very small part of the suspension." In Piccard's opinion, the problem was clearly a design deficiency with the balloon envelope rather than a manufacturing anomaly. "I didn't see anything to indicate that there was anything wrong with any seam."

When the tattered remains of the balloon were recovered from a school playground in east St. Paul and inspected later that day, the Litton team found it impossible to determine precisely what had gone wrong. Postflight inspections of these kinds of plastic balloons are not usually very instructive: the polyethylene is so fragile that it typically develops numerous rips and stresses in free fall following release and deflation. Still, Nick and some members of the Litton crew pointedly disagreed with Piccard's analysis and questioned his design expertise. They were convinced that a seam *had* in fact split when the balloon encountered the six-knot wind shear, most probably a result of defective material. Karl Stefan concurred, believing that a manufacturing defect of some sort was the likeliest culprit, but he was skeptical that wind shear was to blame. He would later point to a study by respected meteorologist Alvin Morris ostensibly proving that even powerful wind shears would not be capable of ripping a well-built balloon in flight.

"It could have been a faulty seam," Stefan said. "It could have been that in packing up or unpacking, something got scratched. But shear wind never seemed likely to me."

The gondola *Frustrations End*—quite inappropriately named, as it turned out—came to earth in the Battle Creek Park neighborhood of Maplewood, just north of St. Paul, where a small crew, after extricating the big parachute from the bare limbs of an oak tree, loaded it onto a Ford pickup for its return trip to the Litton plant.

Piccard would always believe that his instinct had been correct and that Nick should have somehow found the extra money and hired Schjeldahl, though he had no criticism of Litton's performance on launch day. "It was very well organized," he maintains. "Everything went like clockwork. Still, I feel he should have gone down to Northfield. A reinforced Mylar balloon would have worked, and worked the first time. The polyethylene was cheaper, sure, but what did it save him in the end? It was simply the wrong balloon for the job. And they didn't guarantee the balloon. If Litton had guaranteed the thing, fine, but . . ."

Grievances

Nick was typically straightforward and straightforwardly uncowed at the press conference at the Capp Towers later that morning. He was also careful to maintain the image of the intrepid aviator. A *Minneapolis Star* reporter described his theatrical entrance: "Piantanida threw out his hands in a gesture of futility, kissed his dark-haired wife, and lit a cigarette." Of his own disappointment, he confessed: "I couldn't begin to put it into words. My wife and I have planned and sacrificed for two years for this. Physically I feel great. If there was a balloon available, I'd be ready this afternoon to climb right back into that pressure suit and give it another try."

Janice remembers when she first saw Nick following the flight. "He was so disappointed. But by the time I saw him, he was trying to shake it. 'I can't believe it. I just can't believe that happened.' And then, before we got to the press conference, he suddenly leaned over and said, 'Honey, I'm going to go again.' He just decided on the spot. And that was it." Nick's resolve to carry on could have surprised no one, least of all Janice.

When a *Minneapolis Tribune* reporter asked for an explanation of the balloon failure, Nick was generous in his assessment of Litton's performance. While he said it was "much too early" to determine the cause of the failure, he allowed that the plastic envelope's rupture was "not altogether unexpected." He continued: "All balloons, whoever the manufacturer, have a certain percentage of failures. It's just unfortunate it happened this way." In an aside, however, he told one reporter, "I'm bitter, a little bit annoyed, and frustrated."

Jacques Istel followed Nick with a hastily prepared announcement that Strato-Jump had reached an agreement with Litton for another balloon and promised that a second attempt would again originate from Minneapolis–St. Paul. "Future plans are for another launch as soon as possible," Istel said, assuring all interested parties that Strato-Jump had the funds in hand for a repeat attempt. "We're going back to the drawing board."

"We find it very pleasant here," Nick chimed in. "Except perhaps from 22,000 to 23,000 feet." When pressed for a date, Nick guessed that he would be back in the gondola within four to six weeks.

After the press conference, an unnamed Litton official volunteered some off-the-record speculation about the cause of the balloon failure that seemed to point a finger back at Litton's own balloon operations. "A number of things could have caused it to burst," he admitted, "including rapid expansion of the helium due to the quick ascent." More likely explanations involved the load tapes or the plastic itself. Litton and other plastic balloon manufacturers had experienced their share balloon bursts in the 30,000- to 35,000-foot region (which Litton had noted in its original proposal to SPACE, Inc. that summer) in the months and years prior to Strato-Jump, but new plastic films, new manufacturing techniques, and improved designs had given the company a renewed confidence in their envelopes. Nevertheless, the brittle properties of polyethylene when exposed to extreme cold were still a serious problem for balloon manufacturers in the mid-1960s.

While all parties (Litton, ONR, the Marines, the FAA) praised Nick's nerveless efficiency in handling the emergency, the good face he put on for the press didn't reflect his true feelings about Litton Industries and the job they had done. Privately, he was furious. And with some justification. In spite of the fragility of frozen polyethylene and radical wind shifts at the tropopause, postlaunch balloon failures on manned flights like this were actually quite rare in postwar American aeronautics. Not a single one of the many super-high-altitude balloons built for the Air Force and the Navy by Winzen Research, for example, had failed in flight with a man on board. The one exception may have been the Navy's Strato-Lab flight of 1956. Concerning an unusually abrupt descent from the flight's peak altitude of 76,000 feet, the project's technical report stated, "The flight was prematurely terminated by a malfunctioning balloon valve." The valve in question was likely the valving appendix on the balloon itself and not the electromechanical valve at the top of the envelope. If, as some experts believe, the appendix became pinched, preventing free gas from escaping the balloon, the appendix may in fact have developed a rupture.

A few days after filing the required accident report with the FAA, Nick wrote a long memo to Jacques Istel. In it he listed twelve areas in which he believed Litton had erred, unfairly taken advantage of, or misrepresented themselves to Strato-Jump and SPACE, Inc. In this cornucopia of grievances Nick made the following charges:

1. Litton misrepresented time spent on the project and subsequently overbilled for its services.
2. Litton neglected to supply accurate accounting information as the project progressed.
3. Litton had run into unexpected difficulty attaching the end fitting to the top of the balloon, indicating a possible manufacturing defect.
4. Litton personnel stripped and resealed seams haphazardly during the final stages of balloon construction.
5. Litton personnel held intense last-minute inspections followed by suspicious and private ad hoc conferences prior to balloon inflation.
6. Even though the existence of six-knot wind shears at 23,000 feet was known prior to launch, appropriate warnings were never issued to the Strato-Jump crew or to the pilot.

7. In spite of the 700-800 foot-per-minute ascent rate specified in the flight plan, the balloon actually rose at a rate as high as 1,300 feet per minute as determined by ground-control radio and verified by altitude-versus-time calculations following the flight.
8. Since the ascent rate for the flight could be influenced primarily by only two factors (adjustment of the load or adjustment of the volume of lifting gas)—and since the load had *not* been adjusted—the launch crew likely overinflated the balloon.
9. Based on overheard conversations, Nick believed that Litton officials were aware of the overinflation prior to launch and—for whatever reasons—neglected to share the information with anyone from Strato-Jump.
10. The balloon most likely contained a faulty seam since balloons such as the one Litton built for Strato-Jump are typically designed to withstand twenty- to twenty-five-knot wind shears.
11. Litton deliberately misled Strato-Jump about the company's intention to participate in a follow-up flight in order to avoid negative publicity and decided even before the aborted flight to have nothing further to do with Nick.
12. Litton public relations personnel were rude and disrespectful to reporters and photographers, as well as to Nick and Jacques Istel.

Regardless of the validity of Nick's charges, Litton executives were steaming mad. They had opposed the Strato-Jump flight from the beginning, worrying about liability and bad publicity, and now their name was being dragged through the newspaper mud as the organization that had deposited a potential American hero in the St. Paul city dump. And so Karl Stefan, who had been the point man for the project within the corporation and who had argued for the potential public relations benefits of an association with Nick Piantanida, took the fall. Stefan was fired a few days later.

About a month after the failed flight, Litton refunded the unused balance (a mere $603) of SPACE, Inc.'s deposit, thus effectively ending the relationship between Litton Industries and Project Strato-Jump. It wouldn't be long before Litton's contract with the Air Force Cambridge Research Laboratory would expire and the entire balloon group would be shut down for good. Nick, on the other hand, still had the support of the David Clark Company and would continue to have the use of its pressure suit, but he needed a new balloon if he wanted to make another attempt on the world record.

Nick followed up the October attempt by formally thanking all of the key participants—including several members of the Litton team—in writing. Since he was determined to organize another Strato-Jump flight, it was important that the individuals and support organizations who had backed him in Minneapolis remain motivated to do the same again. He would also be obliged to spend a fair amount of time and energy revisiting his funding sources and reassuring them that not only was the project still very much alive, but that he was personally as committed as ever. At the same time, he wanted his benefactors to know that SPACE, Inc. had legitimate ambitions beyond Strato-Jump. In fact, Nick was already beginning to talk publicly about future efforts that would succeed the world-record free fall. The first follow-up to Strato-Jump, he revealed to a Brick Town reporter, would be an attempt on the altitude record for a night jump. The night-jump record stood at about 40,000 feet, and he vowed to best that by at least 20,000 feet in the spring of 1966.

As he worked on plans for a second attempt at the free-fall record, plotting his next moves, Nick was an active presence at the Lakewood drop zone that autumn, making sixteen more jumps before Christmas. In late November Nick and Phil Chiocchio jumped at 5,000 feet over Eatontown, New Jersey, aiming for the Monmouth County Shopping Center where hundreds of children had gathered to witness a holiday miracle. Phil left the Cessna first, dressed as a North Pole elf, with Nick following a few minutes later in full Santa Claus costume. Skydivers of any sort were still very much a novelty, and Phil landed near the crowd of kids to wild cheers. But by the time Nick jumped, the winds had shifted, and he had to track hard to hold his position. When it became apparent that he was in danger of coming down well short of the shopping center, he turned downwind and made a risky, high-speed pass directly over the heads of the crowd. This necessitated a final sweeping turn back into the wind only feet above the shopping-center parking lot. As Nick completed the turn, his Para-Commander oscillated slightly and stalled. The holiday crowd watched in horror as Santa Claus slammed hard sideways into the asphalt and was dragged across the parking lot by the still-inflated parachute.

One of the kids, with tears suddenly filling his eyes, turned to Santa's elf and asked, "Is Santa all right?"

"Oh, sure," Phil told the little boy, hoping he was right. He wondered if Santa Claus had ever made his Christmas morning rounds from an ambulance.

Nick managed to get the air dumped out of the canopy and scramble to his feet, still within sight. He immediately jumped onto the hood of a nearby pickup truck, straightened up his glue-on beard, and waved triumphantly at the relieved crowd, yelling, "Ho, ho, ho!"

It was just one more story for the drop zone: the day Santa Claus almost bought the farm.

In spite of the failed Minnesota flight, Strato-Jump was solvent. Though Litton had refused to guarantee the performance of its polyethylene envelope, Istel had insured both the balloon and the gondola with Marsh and McLennan, so funds for a replacement envelope would not be an issue. In addition, exclusive rights to Mahon's prospective footage of the big jump had been sold to Gay Robert, Inc. for 30 percent of the gross proceeds, of which $10,000 was to be paid to the project immediately upon any balloon with Nick aboard reaching 85,100 feet; a jump from that height would allow them to surpass the Russian Andreev's free-fall mark. *Life* magazine had already paid $5,000 for exclusive periodical rights, excepting the first-person story Nick had agreed to write for *True*. Pioneer Parachutes indicated that it planned to continue with its generous donation of equipment, and the Chemical Bank New York Trust Company had loaned the project another $20,000 in cash.

Ironically, Nick's personal financial situation was as grim as ever. The future windfall from Strato-Jump was still as much of a fantasy as had been the diamonds of Venezuela. He had to borrow $500 from SPACE, Inc. in November to cover back rent and to pay off some earlier loans. Janice had recently given birth to their third daughter, Debbie, and without the regular income from Nick's trucking job times in Brick Town were tough.

In spite of it all, Nick's focus was unwavering. There was little time to fret over personal matters, and—even though the first phase of Strato-Jump was finally over—even less time for basketball heroics, the Ichi-Ban, or the streets of Union City. Nick hunkered down in Brick Town and thought about little else but the project. With an expensive failure now under his belt, the stakes had been raised another notch.

There was, of course, one pressing order of business. For the project to retain any semblance of credibility, it needed a new balloon and launch contractor. And it needed them fast.

4 STRATO-JUMP II

Second Chance

Nick moved quickly and decisively. Even as he faced the Minneapolis reporters in late October and defended Litton Industries against the suggestion that the balloon failure was somehow a result of the company's manufacturing deficiencies or—worse yet—negligence, he had already begun to inquire about other balloon builders. Raven Industries had been founded by veterans of the same elite General Mills balloon division that had been acquired by Litton. But in important ways Raven had made itself into the more innovative and successful concern. Nick had been suitably impressed with the facilities in Sioux Falls when he had visited with Don Piccard earlier in the year, and he decided now that Raven represented his best shot at the stratosphere—even though the company had precious little experience with manned high-altitude flights.

Nick phoned the Raven offices on November 2, just two weeks after the Pig's Eye Dump disaster, and spoke with Russ Pohl, the assistant chief engineer. Following a quick call to an individual he knew at Litton to verify that SPACE, Inc. had paid its bills, Pohl agreed to set up a meeting with Nick that Friday, November 5. The meeting went well, Nick made a good impression on Raven's engineering and operations organizations, and the two sides agreed in principle to a working relationship that would encompass not only Project Strato-Jump but, if all went well, future ventures as well.

Now that Nick had been through the drill, his confidence in his ability to manage a high-altitude operation was sky-high and he headed into Sioux Falls that winter knowing exactly what he wanted and exactly how to get it. Certainly the change in company culture was dramatic. The Litton operation, in spite of the personal attention of such dedicated individuals as Tim O'Malley and Karl Stefan, had the hard-edged, big-city, big-company feel of a diversified corporation; Raven, in sleepy, frozen Sioux Falls, was more like an extended family. As the firm's meteorologist put it: "Raven Industries at that time was a very young and very alive company. We lived and breathed ideas and exuded the energy to make them materialize."

One of the first members of the Raven family Nick got to know was cofounder Paul "Ed" Yost, a stocky, salt-of-the-earth Midwesterner with a brusque manner and an irreverent, hands-on approach to aeronautical engineering. Yost was a blue-jeans, beer-drinking guy with a broad face, an imperial Roman nose, and a shock of wind-fanned hair. He was a gifted storyteller with a rich sense of humor—one of the Strato-Jump crew members described him as a "Will Rogers character"—but he was never the easiest man in the world to work with. His instinctive style of problem-solving often clashed with the by-the-book methodologies of his engineering colleagues, and he could occasionally be stubborn and dictatorial. On the other hand, he was indisputably one of the best scientific balloon flight operations managers around and would eventually become one of the greatest balloonists and balloon builders of the modern era. In later years he would personally break most of the existing world distance and endurance records for balloon flight.

Yost had graduated from the Boeing School of Aeronautics in Oakland, California, in 1940 and completed his cadet pilot training with the U.S. Army Air Corps during World War II. He spent the next several years flying for commercial airlines and as a bush pilot in remote corners of Alaska. In 1950 he became an associate engineer and tracking pilot at the Aeronautical Research Lab, Mechanical Division, of General Mills in Minneapolis. In its heyday General Mills had been a great innovator, and in 1952 Yost participated in a landmark program that sent massive plastic balloons into the stratosphere for the purpose of gathering data on cosmic radiation.

Ed Yost: the Will Rogers of Raven Industries. (Courtesy Joel Strasser)

In 1956—at about the time Nick was hoisting himself onto the top of Devil Mountain in Venezuela—Yost and three others left General Mills to found Raven Industries. One of the company's first commissions was from the Office of Naval Research. ONR wanted a lighter-than-air vehicle capable of carrying a man and enough fuel for three hours of flight at altitudes up to 10,000 feet, with further requirements that the system be small, lightweight, reusable, and capable of rapid launch.

After experimenting with plastic films and portable heating systems—plumbers' pots full of kerosene at first—Yost, with the help of others at Raven, came up with a breakthrough design that would in short order give birth to the colorful phenomenon of recreational ballooning. Using a nylon envelope and a novel propane burner, Yost launched from a deserted airfield near Bruning, Nebraska, on October 22, 1960. He was aloft for twenty-five minutes and landed three miles from his takeoff point. It was the world's first free flight in the modern hot-air balloon, the Kitty Hawk of lighter-than-air sport flying. Three weeks later, with a larger burner and a more capable envelope design, Yost made a flight lasting nearly two hours and reaching an altitude of 9,000 feet.

Raven sold its first sport balloon, based on Yost's design, in 1961. Two years later—and two years before Nick Piantanida showed up in Sioux Falls—Yost and Don Piccard had proven the robustness of Yost's system by making the first hot-air crossing of the English Channel, from Rye, Sussex, to Gravelines Nord in three hours and seventeen minutes in a 60,000-cubic-foot Raven balloon they called the *Channel Champ*. In 1969 Yost would cofound the revival of the Balloon Federation of America (an early sixties gas-balloon organization of the same name had dissolved). The reincarnated BFA was the first national organization for the increasingly popular sport of hot-air ballooning.

But beyond his own expertise and experience with lighter-than-air flight, perhaps the best argument in Yost's favor from Project Strato-Jump's vantage point: he had a fierce reputation for safety consciousness. Nick Piantanida's cocky confidence and Ed Yost's hard-boiled pragmatism sparked an immediate and mutual respect. The two men found that they liked each other immensely.

So, on November 15, at Nick's request, Raven Industries submitted a formal proposal and bid to SPACE, Inc. The proposal offered a couple of options for the balloon. For $4,000 Raven could build a 4,000,000-cubic-foot envelope (just a little bigger than the Litton balloon) that would have a ceiling of 118,000 feet. Or, for an additional $400, they could offer a 5,000,000-cubic-foot envelope that would give Nick a ceiling in the neighborhood of 123,000 feet. Because Raven had already launched five of the larger envelopes without a single failure, and because the factory was already tooled up to build them, they urged Nick to consider the larger balloon.

Additional expenses for the project would involve the rental of instrumentation and rigging ($1,375), materials and expenses for launch services, which included helium and both air- and ground-support vehicles ($2,340), and the labor of Raven personnel ($6,109). Added to the cost of the larger balloon, the bill would come to approximately $14,624—about $5,000 less than Litton's final invoice. It was a good deal.

Upon receiving the bid, Jacques Istel immediately countered with a proposition. He suggested that Raven contribute between $15,000 and $20,000 to Project Strato-Jump (in effect, a donation of time and materials) in return for 16 percent of SPACE, Inc.'s outstanding common stock. Istel painted a rosy picture of the future that would likely follow in the wake of a successful world-record jump.

"It appears that if the first jump is successful," Istel told them, "substantial income may be realized from film rights on future projects while awaiting government contracts." Raven had no interest in Istel's deal, but the offer demonstrates the creativity with which Istel was working the finances following the first flight. The total cost of equipment and services for a second attempt would be higher, in the neighborhood of $120,000, but the cash outlay for SPACE, Inc. appeared manageable. With the project now on a relatively sound, if still somewhat meager, financial footing, the focus had begun to swing from present circumstances to future possibilities. Strato-Jump, then, blew into South Dakota with its own not-inconsiderable momentum and optimism. Certainly in Nick's view, given his embarrassment at the premature ending of the Minnesota flight, things were once again looking up.

The upper Midwest is a place of unique importance in the history of American high-altitude ballooning. Most of the United States' great scientific research flights of the 1920s and '30s had been launched from South Dakota, Minnesota, or Illinois. Factors that had favored this region of the country included good wind-protected launch sites (such as the historic Stratobowl near Rapid City, South Dakota) and the high concentration of primary cosmic rays, which were the subject of much research, in the skies above it. After the war, the appearance of the General Mills balloon operation and the University of Minnesota's research labs would further enhance the area's attractiveness for high-altitude aeronautics. Later, General Mills' expertise would be dispersed not only to Litton (as a result of Litton's acquisition of the Mechanical Division, which only incidentally included the balloon group, in 1963) and to Raven, but to their principal competitor throughout the busy decade of the 1950s, Winzen Research. Winzen had won most of the important postwar balloon contracts from the Air Force and Navy, but by the mid-sixties Raven was in ascendance. (Years later, following founder Otto Winzen's death, Raven would acquire the remains of the Winzen operation.)

What Nick Piantanida had come to Raven looking for in the fall of 1965 was not so much a cheaper balloon, but a better balloon, a balloon that would remain intact as it rose through the layers of atmosphere. And a better balloon was precisely what Jim Winker, formerly of General Mills and the Air Force Cambridge Research Center—now Raven's chief engineer—could offer. With degrees in both aeronautical engineering (he had studied under Jean Piccard at the University of Minnesota) and business administration, Winker brought a facile mind and a practical professionalism to the operation. Under Winker's guidance Raven had perfected the use of thinner, stronger films for balloon envelopes. Thinner plastics meant more efficient vehicles capable of lifting heavier payloads. Litton was still using 1.5- and 1.0-mil plastic films at the time of Nick's Minnesota-based Strato-Jump flight. Raven had by that time already transitioned to 0.75-mil polyethylene as their dominant choice, meaning that the balloon Winker could offer Project Strato-Jump was at least as strong as the one that had burst and deposited Nick in the dump—but 25 percent thinner, a material less than half the thickness of the thinnest human hair. Couple a state-of-the-art, tried-and-true balloon design with Ed Yost's considerable flight operations experience, and it is easy to understand the quick decision to relocate Strato-Jump to Sioux Falls.

Not insignificantly, Winker and Yost, who had worked together so successfully on the first hot-air design, also held the world record for the highest unmanned balloon flight: a 100-pound payload raised to an altitude just above 150,000 feet, launched from Sioux Falls in September of 1959, an absolute record for any class of balloon at that time.

Ed Yost, a major factor in Nick's enthusiasm for Raven, would become an important player in the drama that was about to unfold. Nick had surrounded himself with an eager posse of New Jersey–based skydivers, but he was plagued by that problem common to all charismatic leaders: everyone was eager to tell the boss what they thought he wanted to hear. Nick's physique, his carriage, and his confidence were in constant conspiracy to silence critics and wise counselors alike. But one thing he could bank on: Ed Yost would never tell you what you wanted to hear. Ed would tell you strictly what you *needed* to hear. If you were smart, you listened. When Ed talked, Nick not only listened, he heard. According to one observer: "If six people were saying to Nick, 'You gotta do this,' he would check with Ed and if Ed agreed, he'd take it seriously." Yost may have been the one individual involved in the project who had the power to change Nick's mind on important issues. It was a good match from the start, especially since Yost quite clearly felt a personal responsibility toward Nick.

Janice had responded immediately to Yost and to the more personal atmosphere at Raven: "Oh, I liked Ed right away. Liked him a lot. He was very loving, very warm. And Charmian, his wife, she was absolutely wonderful to me." Janice was also quick to spot the bond that was developing between her husband and Yost. "They hung around together even when they weren't working on the project. That was unusual for Nick. And it wasn't like he was following Ed around or Ed was following him—they just hit it off. You know, Ed loved what Nick was doing, and he just made us feel like part of the family."

Because Raven Industries served as a contractor on projects such as this—for Strato-Jump, Raven had been hired to provide a balloon and flight services (including launch, communications, and recovery)—there was no particular incentive for Raven employees to become emotionally attached to the project or those involved in it. But Strato-Jump turned out to be different in a number of respects. Not only did the project's mission capture the collective imagination of much of the Raven team (and particularly of Yost), but Nick connected personally with a number of the Raven crew. Ed Owen, Raven's president as well as one of its founders, eagerly welcomed Nick and Strato-Jump, though he would later become exasperated with Nick's habit of unilaterally commandeering Raven facilities for his own ad hoc press conferences. It seems that nearly everyone liked Nick, wanted to please him, certainly wanted to do a good job for him. Jim Winker, not a man to rush to judgment on anything and not inclined to hyperbole, found himself impressed by Piantanida: "I considered him to be qualified in his field, very qualified. He was clearly in my mind *not* in the kook category."

The participation of Yost and Winker ensured that nothing within Raven's purview would be overlooked. Everyone worked hard to reward Nick's choice to bring his project to Sioux Falls. After a meeting with Raven representatives on December 29, the FAA office in Rapid City issued an experimental certificate of airworthiness for the gondola—which Nick had personally picked up from the Litton plant in St. Paul and delivered to Sioux Falls—and a new Raven-built balloon. The agency, of course, was familiar with Raven's reputation. The first requirement listed in the FAA directive concerning the proposed flight: "This balloon shall not be operated in free flight unless under the direct control of Mr. Ed Yost."

Prior to the second Strato-Jump attempt, Nick created a formal project plan, which documented objectives and a training regimen and provided a brief specification covering balloon, gondola, parachute, survival system, communications, and flight operations. Raven was already having a salutary effect, helping to bring a needed discipline and maturity to the activities of Project Strato-Jump.

One of the project's goals was to establish that a trained skydiver could negotiate free fall through the stratosphere without becoming a human centrifuge and succumbing to a potentially fatal flat spin, something that had eluded the Air Force's Project High Dive in the mid-1950s. Some of the dummies the High Dive team had dropped from high-altitude balloons had promptly gone into flat spins at rates as high as 500 revolutions per minute, leading eventually to a search for some sort of stabilization mechanism. Yet, as Nick's new project plan explained, "Present methods of survival, which employ drogue devices, have disadvantages. Proving that stabilizing devices are not necessary when a man is able to use his body as an aerodynamic foil, may greatly simplify current emergency procedures."

Certainly Joe Kittinger's experience on his first Excelsior flight in which a prematurely armed drogue wrapped around his neck and nearly killed him would testify to the potential disadvantage of complex parachute systems. But Excelsior had operated under the assumption that high-altitude aviators (including astronauts) would be skydiving neophytes and would require some form of stabilization for an emergency descent. Nick's project plan addressed this issue:

> We believe that special free fall training should be introduced into all programs involving flight at extreme altitudes. Putting aside the most important issue, a man's life, our government cannot afford to lose its financial investment in the cost to train these men. Each astronaut, for example, represents an investment of 6 and 1/2 million dollars.

Nick and Jacques Istel both liked to refer to Strato-Jump—perhaps a bit disingenuously—as a research project. There was an unspoken assumption that the whole business had to be about more than one man's lonely quest for glory. The original certificate of incorporation for SPACE, Inc. had listed some of the fields to which Strato-Jump hoped to contribute, including aerodynamics, aeronautical engineering, space medicine, and "the sciences of the descent of objects and persons through space, the earth's ionosphere, stratosphere, and atmosphere." Cloaking the project in the mantle of research gave it legitimacy and made it easier to raise money, recruit staff, and to secure test facilities. But, in fact, free fall from stratospheric heights without stabilization had yet to be satisfactorily proven to the American aviation community (despite the success of Andreev in the Soviet Union), and the Strato-Jump team would be able to make a legitimate claim to discovery if they could successfully complete their program. "I've spent two years getting ready for this," Nick insisted to doubters. "Sure I think it has scientific value."

By the time he arrived in Sioux Falls, Nick's skydiving experience was considerable for someone who had taken his maiden jump just a couple of years before. He could already claim 435 jumps, 90 of which were from altitudes above 10,000 feet—making him one of the most experienced high-altitude jumpers in the country. By rough calculation, he had free-fallen a distance equivalent to a trip from Boston to Washington, D.C.

On December 14 Istel wrote to Raven accepting their proposal and agreeing to their financial terms. But in order to avoid the delays the project had experienced with Litton, Istel stipulated that Raven would have to commit to providing the balloon by January 30, 1966, or face late fees of $100 for each additional day.

Shortly after the Litton balloon failure, Dr. Mohler at the FAA had contacted Nick to once again encourage him and to assure him that the agency would continue to support future world-record attempts. If anything, the Minnesota attempt had firmed up the resolve of Nick's support network. "I know this was a real disappointment," Mohler wrote, "but don't give up. Knowing you, I know you won't! Malfunctions were experienced in initial flight tests by Wiley Post and others, so, history is repeating itself." He cautioned the press not to underestimate Strato-Jump: "No one has ever heard of Nick Piantanida, but no one had ever heard of Wiley Post, either." Mohler would be transferred to Washington, D.C., before Nick could launch again, but he promised to speak to his successor in Oklahoma City and to make sure that FAA personnel there would continue to be released from duty as needed to participate in future Strato-Jump activities.

SIOUX FALLS

Nick and his Strato-Jump crew—engineers, technicians, flight surgeons, and official observers—began trickling into Sioux Falls in January 1966, invigorated with a renewed optimism and eager for a second attempt to rewrite the record books. Magazine and newspaper reporters, photographers, and Mahon's motion-picture company were again on hand to document the activities, and this time—everyone hoped—to capture the triumph itself.

Most of the crew, which Nick had requested report for duty no later than January 29, was quartered at the Town House Motel, a Best Western affiliate on Phillips Avenue at the south end of downtown Sioux Falls. In the late 1950s federally funded urban renewal had destroyed much of early industrial Sioux Falls and replaced grubby warehouses and factories with modern buildings that housed banks, offices, and motels like the Town House. Nick had met the motel's owner, twenty-five-year-old Dick Kelly Jr., on an earlier visit and had so affected the younger man that accommodations for Nick and Janice had been provided gratis for the duration.

"I was in awe," Kelly said of his first meeting with Nick. "He was a great big charismatic guy—very impressive." The dark, comfortable lounge at the Town House, called the Black Watch, became a hangout and home base for the crew, but they also got to know some of the other interesting watering holes of Sioux Falls. A popular pastime during the frigid evenings was barhopping in search of the perfect hot-buttered rum. The more adventurous among them also frequented an establishment two blocks away from the Raven plant on the old north end of town. Jim's Barbeque, in a rundown firetrap of a building, was a raucous joint that served some of the best brick-oven spare ribs in the Midwest. It was one of the few downtown businesses in Sioux Falls that was owned by an African American, an elderly former Navy chief who went by the name of Smoky Jim Lee. Jim's suited the rough-and-tumble crowd of parachutists and pilots Nick had assembled. There was a perpetual cloud of greasy hickory smoke floating just above head level, and there was a great jukebox blasting rhythm-and-blues hits. On occasion Smoky Jim was known to break out his private stock of 120-proof rum. Fights were not uncommon.

"It wasn't a place you'd take your family," Dick Kelly recalls. Another occasional visitor noted, "It was a pretty rough joint. You didn't want to go back and look at the kitchen." Directly above the restaurant was another establishment, a remnant of conservative Sioux Falls' raunchier post-frontier days. The Twentieth Century Club was a thriving whorehouse, and a few members of the Strato-Jump crew got to know the way up. From the second-floor windows they would have been able to see the red-brick Raven building where the gondola and balloon were being prepared for what everyone in town was hoping would be the highest balloon flight and the longest parachute jump in history. As a community, Sioux Falls was ready to take its place in the record books.

At the dawn of the twentieth century the altitude record for a manned balloon stood at 35,424 feet, achieved by German meteorologist Arthur Berson and his friend Reinhard Suring. Without better knowledge of high-altitude physics and physiology, and without better survival systems, it would have been difficult to improve on that. Then, in 1927, brash American Army Air Service Capt. Hawthorne Gray, carrying two small cylinders of pressurized oxygen and plenty of blankets in his balloon's wicker gondola, launched from Belleville, Illinois, and reached 44,000 feet—higher than anyone had been in any type of aircraft. Regrettably, in his eagerness to establish a new record, Gray lost track of time and exhausted his oxygen supply before he could descend into sufficiently breathable atmosphere. He was dead by the time he returned to Earth.

The first human beings to reach the stratosphere and live to tell about it were Swiss physicist Auguste Piccard (Don Piccard's uncle) and his assistant, Charles Kipfer, who made a flight in a sealed aluminum sphere to 51,200 feet in 1931 and bested that mark by 2,000 feet the following year. The flights of Gray and Piccard ushered in a hopeful flurry of high-altitude balloon activity that resulted in the first "space race" between the Americans and the Russians. This period ended with the 1935 flight of the American aerostat *Explorer II*, which reached an astonishing height of 72,395 feet, deep into the stratosphere. (Interestingly, seventeen-year-old Ed Yost—while not among the Rapid City, South Dakota, crowd that witnessed the launch of that historic flight—had visited the Stratobowl launch site earlier that year and had followed the flight preparations closely through newspaper and radio, much as future generations would follow NASA's moon shots on television.)

Following World War II high-altitude researchers resumed sending men and balloons into the stratosphere, but with a radical new agenda. A handful of visionaries was beginning to contemplate actual space travel, and the balloon-borne capsule turned out to be a remarkably efficient and convenient vehicle for testing the readiness of human survival systems for the hostile environment of outer space. If a system could be devised to supply a person with oxygen, and to keep that person pressurized and warm at 100,000 feet (above more than 99 percent of the earth's atmosphere), it would in all likelihood be adequate for orbital and even extraorbital flight. Both the Air Force and the Navy funded ambitious manned-balloon programs to probe the fringes of space. And firms like General Mills, Winzen Research, and Raven Industries supplied the means of transport. These postwar years would turn out to be the glory days of American high-altitude aeronautics.

In addition to scientific and sport balloons, Raven had begun some months earlier to work on the world's first known plastic cargo parachute. A prototype was almost ready, and naturally Nick had volunteered to test-jump it. But the lion's share of the work that January involved final preparations for the second Strato-Jump flight. The Raven engineers had made a few suggestions for improvements to the gondola, including the addition of more ballast for a total of fifty-five pounds. Unlike Litton's liquid-ballast system, Raven installed a more traditional, dry-ballast set-up that used fine-gauge steel shot. Raven also suggested spring-loaded mechanisms to ease the opening and closing of the front doors: instead of a single panel that swung open vertically, the gondola would use double French-style doors that swung open horizontally and that were insulated with big waffle-stitched furniture-pad curtains that could be opened and closed independently from the doors themselves. Some new instrumentation would be added for the second flight: a clock would provide the time of day and a pair of gauges would display the air temperature both inside and outside the aircraft. The right-side panel would also contain a timer that required manual resetting every fifteen minutes; this would give both the pilot and the ground a set of temporal benchmarks that would lessen the chance of either becoming obsessed with a problem or set of problems and losing track of elapsed time.

The operational protocol for Strato-Jump II had been changed more profoundly than any of the gondola or parachute hardware. Unlike the arrangements for the Litton flight, the flight controls for Strato-Jump II would reside entirely in the hands of Ed Yost. He would take the balloon up and he would call the shots from the moment of launch until Nick jumped free of the gondola. In the event of a total voice communication loss, Nick would use Morse code to acknowledge the ground every five minutes. A special

The *Second Chance* gondola in the hangar at Joe Foss Field. (Courtesy Joel Strasser)

Morse code sequence was devised for Nick to signal two minutes prior to jump, and another for him to signal an emergency and request immediate cut-down. And while Nick would always have a termination switch available to him, by agreement, Yost would give the order if the decision was made to bring Nick down before he'd left the gondola.

The exterior of the aluminum-framed gondola would be covered with new two-inch-thick panels of molded, bleach-white polystyrene. For the upcoming flight, however, the *Frustrations End* placard had been replaced with a new one that provided an apt name for the spruced-up aircraft: *Second Chance*.

But the biggest single change for Strato-Jump II was in the balloon itself. Nick and Jacques Istel had elected to go with the larger Raven envelope, pushing the upcoming flight's float altitude more than a mile higher than could have been reached with the Litton balloon. The manufacture of the big balloon, which required huge assembly tables more than a third again as long as a football field, was business as usual at Raven where identical balloons had been in production throughout the latter part of 1965. The balloon that would be used on Strato-Jump II (Raven serial number 117) was built on schedule that December and was designed to accommodate five million cubic feet of helium at altitude. The polyethylene was the thinner 0.75-mil type Raven had pioneered. One-hundred-fifty-pound load tapes were employed. The 648-pound balloon would be required to lift a payload of 823 pounds (well within the envelope's limits); the theoretical ceiling for *Second Chance* was calculated to be 124,700 feet.

As January wound down and the early February launch date approached, everyone connected with Strato-Jump—including Sioux Falls locals who had been romanced by the sheer novelty of the project—began to watch the weather. Balloon launches require relatively clear skies and the near absence of wind. Temperature is not a big factor, and this was fortunate because the winter of 1965/66 was a bitterly cold one in the upper Midwest. On January 22 the temperature plunged to an Arctic 28 degrees below zero which broke a thirty-year record for that day in Sioux Falls. A grainy, light snow fell on the region from January 24 through 28, but not enough to cause any concerns for Raven or Strato-Jump. Another cold front moved through on the twenty-ninth, but as the first of February dawned, the skies were clear and the stage was set for the launch of Strato-Jump II.

Raven had complete faith in its resident weather expert, the delightfully proper British meteorologist Emily Frisby, known to acquaintances simply as "Friz." Dr. Frisby had been studying climate and weather since the 1940s with a Ph.D. in climatology from Reading University in England; she had helped develop weather forecasts for the D-day invasion of Normandy and had extensive experience forecasting weather for scientific balloon flights. Having made a number of studies of weather patterns over the American Midwest, she knew as much about the region's stratospheric winds as anyone. In an autobiography published privately in 1987, Frisby referred to the period of her association with Raven Industries as her "balloonatic years" and characterized them as "unquestionably . . . the most vivid of my professional life." Once Frisby gave the all-clear for a launch, both Raven and Strato-Jump personnel knew they could proceed with total confidence.

Thanks largely to the continued efforts of Sen. Harrison Williams, an impressive new contractor was now on board and present for launch preparations in Sioux Falls: the General Electric Company. A detailed project plan submitted in mid-January outlined G.E.'s proposed role. The company's Re-Entry Systems Department, based in Philadelphia and represented by supervising engineer Richard Stiles and a six-man team of engineers and technicians, had agreed to provide velocity measurements for Nick's stratospheric free fall, and to design and fabricate the doppler-type device that would give them accurate readings.

After meeting with Nick and Istel and reviewing the project's preparation and planning documents, Stiles had been suitably impressed. "This is not a bizarre stunt," he announced. "Piantanida has put it on a sound scientific basis."

The tracking system Stiles had in mind consisted of a small transmitter mounted on the rear of the reserve parachute pack. A ground receiver in a van sporting a large single-helix antenna that resembled a prop in a cheap science-fiction movie, perhaps one of Barry Mahon's creations, would pick up the signal from the jumper and give the G.E. specialists the data with which to compute Nick's velocity. All of this hardware was adapted from earlier company-funded space research. G.E.'s engineers had already calculated the maximum free-fall velocity they believed Nick could expect to achieve: Mach 1.07 (just barely supersonic), which could be expected at an altitude somewhere between 87,000 and 90,000 feet.

So while Nick continued to portray the project as part research effort and part glorious adventure—he told reporters his free fall would be "just like floating on a mattress of air" and answered questions about what would happen in the instant he exceeded the speed of sound by saying, "I guess I'll find out when I get there"—Strato-Jump was gradually transforming itself from a rank-amateur, shirtsleeves endeavor into something approaching a professional aerospace program. Nick Piantanida's dream, which he'd once been reluctant to reveal for fear of ridicule, had become a collective obsession that

Nick and Janice share an intimate moment in Sioux Falls during preparations for Strato-Jump II. (Courtesy Joel Strasser)

was quietly but systematically coopting key elements of the American military industrial machine.

Nick had gone home to New Jersey briefly, returning to Sioux Falls with Janice on the third week of January. They had left two of the girls with Janice's parents and the third with a neighbor in Brick Town and headed across the country. They talked about the exultant return trip they would make just a couple of weeks later. It would be the beginning of the end of hard times. Nick would finally have an opportunity to relax with his family and enjoy the fruits of all his labor. They arrived in Sioux Falls and checked into the Town House Motel late on Saturday, January 22.

At the Town House Dick Kelly had fallen completely under the spell of Nick Piantanida. Nick had a quality, when he wanted to exhibit it, of tremendous personal warmth, a magnetism that attracted people to him and that held them in loose orbit, and Kelly had been sucked in. He spent hours sitting in the warm darkness of the Black Watch listening to stories of Angel Falls and of falling through clouds and of basketball games in which Nick had scored more points than the entire opposing team. In spite of the fact that Nick was barely an inch over six feet, Kelly remembers him as much taller, a mistake made by many who encountered Nick. (Don Piccard, who was a full inch taller than Nick, had to consult a photograph of the two of them standing side by side before he would believe that Nick hadn't been a few inches taller.) This can be explained partially by the fact that Nick's long arms, huge hands, and size 13 shoes seemed to belong to a larger man, but the rest of the explanation must be that Nick Piantanida in the flesh was indeed larger than life. Kelly remembers thinking, at some point during those rambling sessions in the Black Watch, that Nick was too much for life.

And then this thought suddenly occurred to him: "Here is a guy destined to kill himself."

In the Tradition

Project Strato-Jump was very much in the tradition of the great postwar high-altitude balloon programs. The Air Force's Project Manhigh made three closed-capsule stratospheric flights: a test flight to 96,000 feet with Joe Kittinger aboard (June 1957), Project Director David Simons's overnight research flight to 101,500 feet (August 1957), and pilot extraordinaire Clifton McClure's near-disastrous ascent to 99,700 feet (October 1958). Manhigh had proved that it was in fact possible to raise a human being to the edge of space and return him safely to Earth and, in so doing, had worked out many of the solutions to hardware design difficulties and procedural problems that would help outfit and prepare the United States for space travel. Yet by the winter of 1958, all the tables had turned. The Soviets' success with their satellite *Sputnik* the previous year had stunned the Western world and forever altered the game, and the United States shifted its focus to rockets and looked to its fledgling civilian space agency, NASA, for redemption.

The Navy's five stratospheric manned balloon flights, known collectively as Project Strato-Lab, could claim both the earliest and the highest postwar stratospheric flights. Strato-Lab made it to 76,000 feet in the fall of 1956 and reached its official world-record of 113,740 feet on the project's swan song in 1961. That last Navy flight ended tragically when Lt. Comdr. Victor Prather, on the verge of exhaustion, slipped into the Gulf of Mexico and drowned just minutes after splashdown.

But it was Joe Kittinger's emergency escape program for the Air Force, Project Excelsior, that had most thoroughly captured Nick's imagination. Kittinger made three flights into the stratosphere in an open gondola and each time free-fell back to Earth. The first Excelsior flight to 76,400 feet in November 1959 had been a nightmare. The timer used to trigger the release of the small drogue chute employed by designer Francis Beaupre's revolutionary multistage parachute system malfunctioned and the drogue deployed prematurely. When the drogue wrapped around his neck, Kittinger entered an accelerating flat spin. He landed unconscious but physically undamaged. The timer-arming mechanism was redesigned, and the flight of *Excelsior II* and associated jump from 74,700 feet went off without a hitch.

The final Project Excelsior flight would become the stuff of legend. Kittinger's right pressure glove malfunctioned during the ascent, but he elected not to inform his ground crew, understanding that they would be obliged to order him to descend. At peak altitude, 102,800 feet—nearly eighteen miles above Earth—his hand had swollen grotesquely. But Kittinger mumbled a prayer and stepped off into history's longest free fall, approaching and possibly even surpassing the speed of sound. This achievement impressed even the most jaded members of the aviation community, including NASA's Mercury astronauts who praised Kittinger and his team for establishing the survivability of superstratospheric bailout. Excelsior offered hope for astronauts who might be forced to abandon a disabled craft in the crucial moments following takeoff. It also heartened pilots of the X-series rocket planes out in the Mojave Desert who were already plying the extreme skies above 100,000 feet.

Kittinger made one more stratospheric balloon flight with the relatively unknown Project Stargazer in December of 1962 as part of a high-altitude astronomy research effort designed to observe the moon and Mars, and it would be the last stratospheric manned balloon flight to be wholly or partially sponsored by any branch of the U.S. military. The money and focus had been almost completely diverted into the race to the moon to

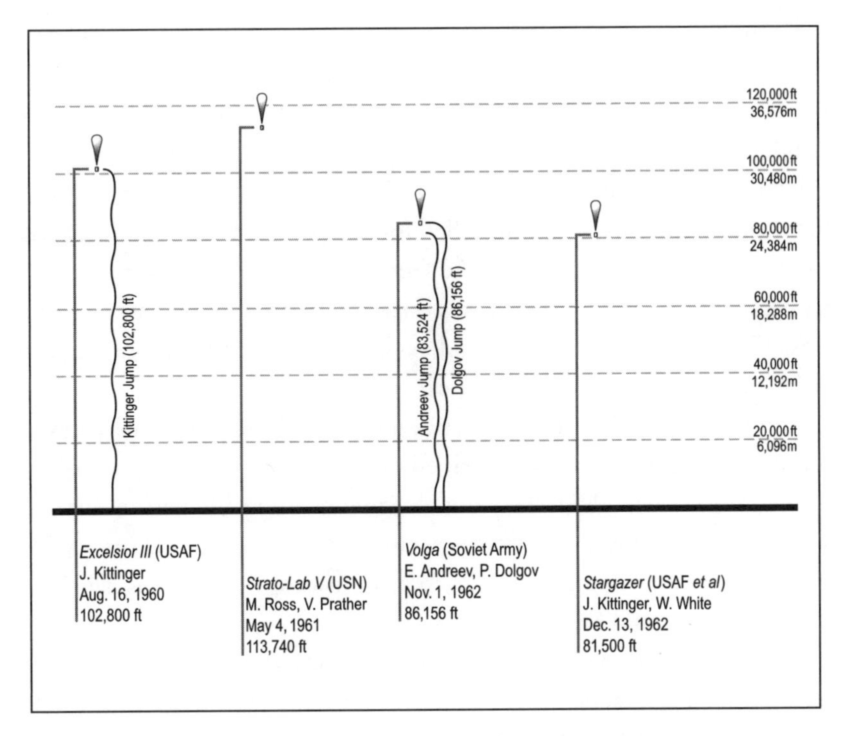

The competition: 1960s high-altitude manned balloon flights prior to Project Strato-Jump. (Courtesy Jonathan Townsend Lee)

which President John Kennedy had so stirringly committed the nation the year before.

When Nick first heard about the free-fall record set by Maj. Eugene Andreev, the available information was sketchy; a critical part of the story of the flight of the *Volga* would not surface until much later. As it turned out, there had been two men aboard the gondola that was launched from a site near Volsk, Russia, on the morning of November 1, 1962: Maj. Eugene Nikolaevich Andreev and Col. Peter Ivanovich Dolgov. Both were expert parachutists with several thousands of jumps between them. Dolgov, a member of the Soviet cosmonaut program who would serve as crew commander for the flight, had been involved in the testing of some of the earliest Soviet space suits and had personally developed the new parachute-opening system that would be used in the jumps from the *Volga*. Andreev was a pilot and a test jumper who held two Soviet records for precision parachute landings as well as a pair of Soviet free-fall records. The two men, along with a two-man back-up crew, made several test flights at relatively low altitudes and spent many hours in decompression chambers in preparation for their stratospheric journey. The plan called for Andreev to jump first and to free fall to a height approximately 3,000 feet above the ground before opening his parachute. His trip down, it was estimated, would take about five minutes. Then Dolgov would jump and—incredibly—open his chute immediately; his descent would require nearly forty minutes.

According to Soviet records, the two men—both wearing partial-pressure suits—entered separate pressurized compartments in the gondola at 7:00 A.M. and spent about forty minutes prebreathing 100 percent oxygen. The *Volga* was airborne at 7:44 A.M. Andreev later recorded his impressions: "As we ascend, the dome of the sky changes its color. At first it is pale blue, then violet blue, and finally black."

At about 10:00, as the balloon passed through 80,000 feet, Dolgov received final clearance for the jumps. Andreev began to decompress his compartment, which caused his suit to inflate, squeezing him so hard that he found it an effort just to breathe. Through the glass wall separating the compartments, Andreev could see Dolgov smiling. Moments later, Andreev pressed a pair of levers that popped the hatch on his compartment and ejected him into the near-vacuum almost fifteen miles above Earth.

Like Kittinger before him, Andreev had the initial sensation of floating rather than falling, even though he would shortly reach a speed of almost 560 miles per hour. Using his skydiving skills to angle away from the great river and its tangle of tributaries, Andreev headed for a giant field. His parachute opened more or less on cue at 3,117 feet and brought him safely to the ground. Andreev quickly spread the big chute out on the dirt so that his comrade could spot it on his own way down. "The whole body aches," Andreev wrote. "I am very tired." He stretched out in the center of his canopy to watch for Dolgov.

News of the senior officer's descent—indeed his very participation in the flight—was initially withheld from the public. Dolgov was later rumored to have been killed either in a launch-pad explosion or—according to one Soviet defector—during a low-altitude test of the Vostok ejection seat. Both rumors were false.

Two years earlier Dolgov had established a world record for the highest "opening without delay" parachute jump: 48,671 feet. Falling with an open parachute from such an altitude is an extraordinarily grueling experience. Rather than a rapid free fall back into warm, breathable atmosphere, a jumper making a no-delay opening must endure the elements for many minutes. One would only undertake such a jump intentionally in order to set a record. Dolgov was attempting nearly to double his own record that morning when he left the *Volga*. He ejected himself from his compartment about three minutes after Andreev's exit as the balloon reached a height of 86,156 feet. He opened his chute immediately, as planned, and the new opening system worked perfectly. It remains the highest personal parachute opening on record.

Yet by the time Andreev reclined on his parachute to watch his comrade's canopy descend, Dolgov was already dead. Investigators later discovered the minuscule hole in the glass of the helmet's face mask that had depressurized the survival system.

Andreev made his jump from an altitude of 83,524 feet and free fell 80,360 feet without a drogue, setting an official world record and proving what Nick Piantanida wanted to prove to the American aviation and aerospace communities: that a trained skydiver can in fact negotiate a stratospheric fall without a stabilization device. Both Andreev and Dolgov were awarded the Order of Lenin and the Golden Star Medal, and both were granted the honorary title of "Hero of the Soviet Union."

If the account of Dolgov's death and its cause had reached the West, and if Nick had learned of it, he might have been at least partially sobered by the very real dangers of the environment he intended to visit.

Or perhaps not. In Nick's own mind, he was still bulletproof. Which may explain Ed Yost's insistence that Nick attempt once more to consult with Joe Kittinger. Yost and Kittinger were good friends, and Yost—with limited first-hand experience with super-high-altitude manned flight—felt that Kittinger's knowledge and appreciation of the dangers involved would be just the prescription for Nick's damn-the-torpedoes approach. Unfortunately for the project, Kittinger was, even with Yost's imprimatur, no more receptive to Nick's latest invitation to serve as a consultant to Strato-Jump than he had been back in 1963. Which did absolutely nothing to discourage Nick.

"I guess I have to go overboard when I try something," he'd told a Minneapolis reporter after the first Strato-Jump flight. "Like this jump. I want to carry some letters to space when I go and mail them later. Maybe I'll mail one to Eugene Andreev."

When Jacques Istel had insisted—prudently—on a life insurance policy for Nick as a condition for his own participation in SPACE, Inc., Nick had become so enraged that Istel had instinctively grabbed a paperweight from his desk in self-defense. "Nothing's going to happen to me, Jacques!" Nick shouted. "Goddammit, I'll have a parachute, won't I?"

Nick's desire to capture the official free-fall record from the Russians could only have intensified in the months since the birth of Strato-Jump. By January of 1966, as the steady buildup of American forces in Vietnam continued, tensions between the United States and the Soviet Union were running high, and the tension was affecting America. In Sioux Falls the Strato-Jump crew could feel the change. Two days before the launch of *Second Chance*, President Johnson had lifted the Christmas bombing halt and ordered his commanders to resume the aerial bombardment of North Vietnam. North Vietnamese air defenses had been equipped with Soviet-supplied antiaircraft artillery, surface-to-air missiles, jet fighters, and radar systems, and Johnson and his advisors chose their bombing targets carefully in an effort to avoid provoking the Soviets or the Chinese into further intervention. The day before the flight, even as the bombs rained, Johnson appeared on national television to announce that he was asking the United Nations to summon North Vietnam to a peace conference.

While ground zero for the Strato-Jump crew and for Raven personnel was the little airport in South Dakota, the increasingly dangerous events in Southeast Asia may have provided a real sense of mission for many of the participants, including Nick Piantanida himself. There would be a special sense of accomplishment when the world record could be claimed by an American, and a sense of serendipity that the battle would be waged and won in the blue skies above the American heartland.

Protecting the Pilot

In the frigid days preceding the launch, the eye of the project's hurricane was the flight operations area in a back section of the Raven Industries building on East Sixth Street, a hulking, five-story brick structure that had been a cookie factory before Raven bought it in 1961 and later relocated from their original home in a wooden Air Force barracks at the Sioux Falls Municipal Airport. As launch day approached, the gondola and the balloon—carefully folded and packed into an insulated wooden crate to protect it from wear and tear—were trucked over to a National Guard hangar at the airport, officially named Joe Foss Field in 1955 after the great World War II ace who served two terms as South Dakota's governor and would later become the first commissioner of the American Football League.

Communications for the flight would be handled both from a station at the airport and—because Dr. Frisby was predicting that prevailing winds aloft would carry the balloon rapidly eastward—from a remote post to be established in the little town of Estherville, Iowa, some 100 miles due east of Sioux Falls. Ed Yost, Jacques Istel, and others would head east by plane shortly after the launch and join Raven electronics technician Darrel Rupp at the Estherville airport. Janice had decided that she, too, would follow the flight from Estherville in order to be nearer Nick's landing site.

Of all the prelaunch work, nothing was so important as the preparation of the survival system designed to protect Nick from the near-vacuum, extreme temperatures, and solar exposure he would encounter in the stratosphere. Until the very end of the nineteenth century, the upper regions of the atmosphere had remained mostly a mystery. It took experiments with sounding balloons to finally establish the modern model of atmospheric layers. The sheath of warm, breathable gas we consider Earth's atmosphere, the troposphere, extends upward to an altitude of between six and eight miles. The *tropopause* is the transition from the troposphere to the stratosphere, and above that point the environment is unforgiving. As one moves upward through the atmospheric layers, air pressure decreases at a rate of approximately 20 percent per mile. By the late 1940s physicists at the Air Force's Aeromedical Laboratory had concluded that, from a human factors perspective, the upper stratosphere is the de facto equivalent of deep space. The Armstrong Line (named for Air Force Col. Harry Armstrong, who had run the School of Aviation Medicine at Randolph Air Force Base and later became the second surgeon general of the Air Force) was said to exist at 63,000 feet. Above the Armstrong Line, according to Armstrong's definitive work, *Aerospace Medicine*, published in 1938, expanding gases will "boil" the blood of an unprotected human being. In the upper stratosphere, bod-

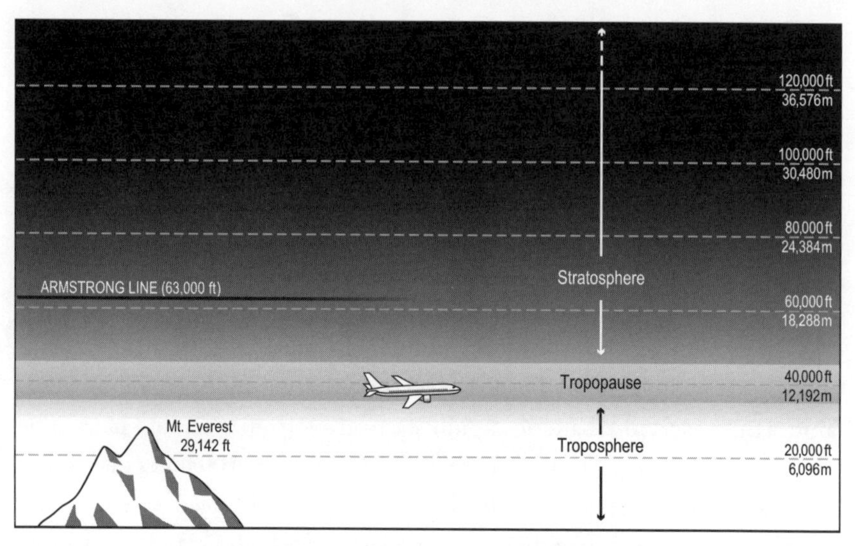

The arena: atmospheric layers. (Courtesy Jonathan Townsend Lee)

ily fluids begin to evaporate at 32 degrees Fahrenheit, the body swells from gas expansion, water vapor fills the lungs, and vital blood vessels rupture. Incredibly, Nick intended to ascend another 60,000 feet *beyond* the Armstrong Line, to the very precipice of true space.

While an unprotected body exposed to a partial vacuum will deteriorate in fairly short order, the effects of a very brief exposure are sometimes exaggerated. As divers know, attempting to hold one's breath while ascending from a higher-pressure environment to a lower-pressure one can damage the lungs and the brain, but a rapid exposure to a lack of air pressure is quite survivable. It is a myth that the body will immediately explode. A human exposed to a near-vacuum has on the order of ten seconds of useful consciousness at his disposal.

After a ten-second exposure a subject will begin to suffer from acute pain in the joints (the "bends") and may experience some minor swelling of the skin and underlying tissues. But because human skin is extremely elastic and relatively strong, the swelling subsides if pressure is restored, and there may be no permanent damage. Shortly before Project Strato-Jump's first flight, a test subject in a decompression chamber at NASA's Manned Spacecraft Center in Houston had been accidentally exposed to a near-vacuum. The subject remained conscious for fourteen seconds, approximately the length of time required for oxygen-starved blood to travel from the lungs to the brain. Fortunately, technicians were able to restore pressure promptly and the subject suffered no physiological damage.

In 1959 a Marine lieutenant colonel named William Rankin survived a bailout without a pressure suit at 50,000 feet—well below the Armstrong Line, but high enough to illustrate the dangers of unprotected exposure to the stratospheric environment. Rankin later described the "savage pain" he had endured: bleeding from all bodily orifices, cramping that felt like the stabbing of knives, inner organs bloating and distorting, eyes that felt as if they were being clawed from their sockets, and thunderous explosions in his ears. Another thousand feet higher, flight surgeons surmised, and Rankin could not have survived the ordeal.

Joe Kittinger's right hand had remained unpressurized during the entire flight of Excelsior III, yet he endured the ascent and the fall back to Earth. In spite of painful swelling at altitude, the hand had returned to normal shortly after Kittinger arrived back on the ground.

As time elapses, however, the problems of exposure to an extreme low-pressure environment accumulate rapidly. After about three minutes irreversible brain damage occurs. Studies with chimpanzees have shown that exposure of more than a few minutes is likely to be fatal. For obvious reasons, the medical community has precious little first-hand data on the longer-term effects on human subjects.

On his trip from Sioux Falls into the surreal region above 100,000 feet where the skies are black at noon and where the earth reveals itself as a giant sphere, Nick would be protected only by the Dave Clark Special which would surround his body with the air-pressure equivalent of about 35,000 feet above sea level. The suit was, in essence, a small, man-shaped space capsule. During ascent gases would inflate the suit and gloves that would enclose Nick to compensate for the rapidly diminishing air pressure. A David Clark technician and the Air Force's Marvin McCall repeatedly tested the suit at Raven, inflating it and deflating it. The helmet, arguably the most critical element of the survival system, would be pressurized and flooded with oxygen. The helmet also underwent repeated inspections at Raven.

On launch morning, it was planned, Nick would participate in a prelaunch ritual that had become standard for the high-altitude balloonists of the 1950s:

prebreathing. Following World War II an important discovery was made. If high-altitude pilots were to breathe pure oxygen for some extended period prior to flight, the nitrogen in solution in their bloodstreams and tissues could be effectively flushed from their bodies. (Earth's atmosphere consists of 78 percent nitrogen and only 21 percent oxygen.) Dr. John Paul Stapp, under whose authority the Air Force's Manhigh and Excelsior projects had been carried out, conducted extensive prebreathing tests in 1946 and 1947. Stapp and his team of volunteer subjects established that prebreathing for thirty minutes immediately prior to a high-altitude flight could reduce, if not eliminate, decompression sickness (DCS). The joint pain associated with "the bends" accounts for nearly 70 percent of all altitude DCS cases.

Without the pure oxygen prior to flight, a pilot ascending above 35,000 feet was vulnerable to cramping and severe joint pain (shoulder pain being the most common manifestation), along with a rarer phenomenon Stapp called "the chokes"—a burning sensation in the trachea caused by nitrogen bubbles trapped in the respiratory system. Later studies would find that a number of factors increase the risk of DCS, including higher altitudes, a faster rate of ascent, a longer exposure to a low-pressure environment, and colder temperatures.

First documented in the mid-nineteenth century by engineers monitoring painful muscle cramps suffered by coal miners in air-pressurized mineshafts, DCS became a serious problem for stratospheric aeronauts in the 1930s. Prebreathing of oxygen, Stapp observed, along with a special low-residue diet, reduced internal gas expansion and eliminated symptoms of DCS. Anyone spending significant time at an altitude of 35,000 feet—the equivalent of which Nick, ensconced in the Dave Clark Special, intended to inhabit for at least a couple of hours—needed to prebreathe oxygen in order to avoid debilitating illness. The prebreathing process continues to be employed today by space-shuttle crews prior to extravehicular activity. Nick was, of course, familiar with the ritual, having prebreathed at both Tyndall and the FAA facilities in Oklahoma, as well as prior to his Minnesota flight, and he was comfortable with the process.

Because Strato-Jump's was an open-gondola aircraft offering little more protection from the elements than the wicker basket of a hot-air sport balloon, the survival system would have to operate flawlessly. On the Air Force's Manhigh flights, the pressure suit and helmet had essentially been failsafe mechanisms in the event of a problem with the pressurized capsule. But on projects like Joe Kittinger's Excelsior and Strato-Jump, the suit and helmet were all that protected the pilot from explosive decompression above the Armstrong Line. If anything went wrong at the altitude Strato-Jump had targeted, Nick Piantanida would be—as Kittinger had so brutally put it before his own landmark flight several years before—"one dead son of a bitch."

For this reason, SPACE, Inc. had arranged for two separate pressure-suit technicians and an engineer to attend to the oxygen system. Richard "Dick" Sears from the David Clark Company would work with Marvin McCall, on another leave from the Air Force. In addition, Earl Clifford—an engineer from the Firewel Company, which had supplied the oxygen apparatus—would be on hand to supervise that aspect of the operation. Project Strato-Jump had lined up some of the most capable contractors and consultants available and, as the calendar turned to February, everyone—the Strato-Jump team as well as the Raven technicians and engineers—readied for what they all hoped and believed would be a trip straight into the record books.

To the Top of the Sky

Late on the night of January 31 Dr. Frisby put a hold on the next morning's launch due to forecasts of moderate ground winds. Everyone remained on standby. February 2 dawned cold: 13 below zero with a three-knot wind and heavy, swirling ground fog. But as the fog lifted and the wind settled, Frisby finally issued the clearance for launch.

Following a final equipment check that was witnessed by FAA officials, the Raven crew began rolling out the 324-foot balloon (lighter but a good deal larger than the Litton envelope) along a long canvas carpet that would protect it from the frozen runway at Joe Foss Field. Jacques Istel had convinced the U.S. Navy to donate the helium supply for the flight, and their helium trailers moved in now and inflation hoses were attached. Approximately 28,000 cubic feet of helium would be introduced into the balloon envelope, a volume that would expand to five million cubic feet at peak altitude. The *Second Chance* gondola was transported on the Raven launch vehicle from the National Guard hangar to the launch spot.

Nick and crew, meanwhile, had used the National Guard administration building to complete the carefully scripted ritual of suiting him up for the

Balloon inflation begins in snowy Sioux Falls; cargo parachute in foreground. (Courtesy Joel Strasser)

stratosphere: insulated underwear, cotton socks, electrically heated socks, boots with exposure covers, electrically heated gloves, pressure gloves, exposure mitts, pressure suit, coverall. A small tape recorder for capturing Nick's free-fall monologue was secured to his right leg and a pack of cover letters was sealed in a pocket on his left leg. The suit also held batteries, a stopwatch, mirrors, cameras, and a barograph. Two small cylinders designated "walking-around bottles" supplied Nick with the 100 percent oxygen for prebreathing during and after the suit-up phase. The prebreathing oxygen was delivered through a facemask Marvin McCall had fashioned for Nick at Tyndall. And once the pressure suit was on, a portable ventilation system—supplied with liquid oxygen by the local National

Raven's inflation truck in position as helium bubble expands.
(Courtesy Joel Strasser)

Guard—allowed Nick to avoid overheating as he moved around the dressing area.

The suit-up was followed by a chute-up. The parachute crew, headed now by Pioneer Parachutes' Bill Jolly, attended to the entire jump rig and saw to it that every aspect of the system was attached, secured, and properly configured. The two bailout oxygen bottles were stowed in the backpack with the main Para-Commander canopy and the drogue. The free-fall velocity recording equipment provided by General Electric was secured between the reserve pack and Nick's chest, and the connections were double-checked. Jolly had arrived in Sioux Falls on January 28 and had spent the afternoon at the Raven plant familiarizing himself with Nick's rig and with the General

Bill Jolly assists Nick with suit-up dry run prior to Strato-Jump II. (Courtesy Joel Strasser)

Electric radio monitoring gear. On Saturday morning, at Jolly's suggestion, Nick had made a test jump from 15,000 feet—which gave him a seventy-five-second free fall—using a standard Para-Commander pack and harness, but with the actual canopy that would be used on Strato-Jump II. In spite of frigid conditions (minus 26 degrees Fahrenheit on the ground), the canopy had performed flawlessly. The next day, Sunday, Jolly and the G.E. team had fabricated new containers to encase the radio equipment Nick would carry. By launch morning, Jolly had gotten himself fully up to speed and was confident that the parachute system would be up to the challenge.

Fully laden, Nick weighed 331 pounds as he was transported from the hangar out to the launch spot on the open tailgate of a slow-rolling station-wagon. He carried 250 commemorative cover letters, most of which contained a form address from Nick congratulating and thanking everyone involved. He intended to distribute them as souvenirs (via surface mail) following the flight. The envelopes bore the legend:

Commemorating Project Strato-Jump
Launched from Souix [Nick habitually misspelled "Sioux"] Falls, South Dakota
flown by
Nicholas J. Piantanida
In an Attempt to Improve
High Altitude Bail-Out Survival Methods
and in Doing So to Surpass
The World's Free-Fall Parachute Altitude Record
The Speed of Sound Without a Vehicle
The World's High Altitude Manned Balloon Record

—signed, Nicholas J. Piantanida
(This cover has traveled via balloon, free fall, parachute, and U.S. Mail).

But two of the envelopes he left unsealed until the morning of the flight so that he could hand-write letters to Janice and to Donna (on behalf of the two younger girls) while he waited for the launch. Janice's was to be handed to her shortly after Nick landed, and he wrote that one first:

My Dearest Darling,

Now the tensions are over—the roughest crises of your life, thus far, complete. You came through like a champ. For me this isn't bad—I'm doing it, and love it. Many people have commented on my courage. This, I feel, has been overestimated—I understand what I'm doing, and without this understanding, I wouldn't even put a parachute on, let alone undertake this project.

The real courage, my love, is yours. Complaining little and sacrificing so much. I love you more than anything. You, sweetheart, are the best thing that ever happened to me. I thank God for being fortunate enough to find and have you.

All my love,

Nick

Nick is delivered to the launch site; Jacques Istel at right.
(Courtesy Joel Strasser)

Nick had instructed Janice that the letter to Donna should be held for her until her eighteenth birthday in 1981:

> Dearest Donna—the biggest of my little sweethearts,
>
> By the time you read this, many years will have past. You won't even have any recolection of what went on during this time. You'll know because of being told, but thats all. You won't remember how you would argue with mommy to wait up for me to come home at night—how no matter what time we would put you to bed you would beg, "Just a few more minutes please."

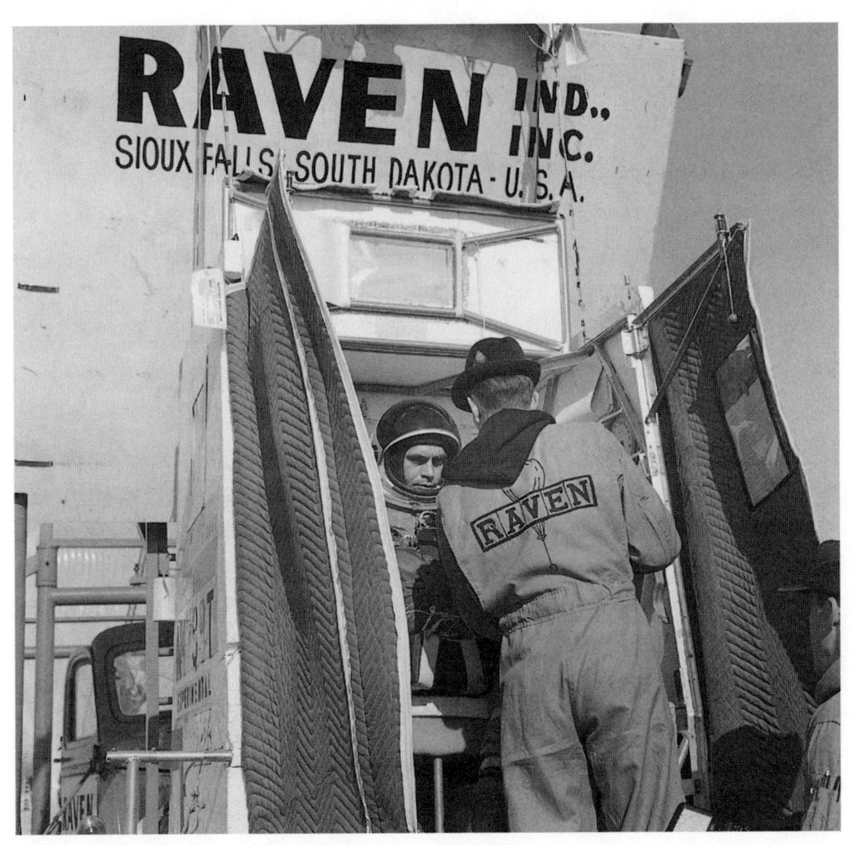

Nick takes his seat in the gondola, assisted by Darrel Rupp.
(Courtesy Joel Strasser))

How you loved to go to church with me, and how you wanted to parachute with me. But—the dearest thing you would do was to come over to me—crawl up on my lap—kiss me on the cheek and say—"I love you lately daddy."

By now you will have loved, and hated, your parents, the "old fashioned" ones, many times. I know, I went through the same crisis many times before reaching 18.

The challenge of life is ahead of you—make the most of it. Walk in the shadow of God, and you shall.

—With all my love your thankful father:

Thankful for being blessed with three wonderful, beautiful daughters.

Strato-Jump II is airborne. (Courtesy Joel Strasser)

He climbed into the gondola for a final communications and instrumentation check. At 11:51 he sealed the faceplate on his helmet and began to breathe pure oxygen from the gondola's main supply. A few minutes after noon, the safety belt across the doorway was latched and the curtained doors on the front of the gondola were closed.

The balloon was released cleanly into a light breeze at eleven minutes after noon as a small crowd of onlookers—reporters, photographers, state police, a few of the Sioux Falls curious who had seen their share of balloon launches over the years and who were not easily impressed—stomped the ground and swung their arms to stay warm. A few remained in their cars with the heaters on, the exhaust forming clouds on the edge of the runway. The fog had dissipated and the snow blanketing everything beyond the cleared

Onlookers at Joe Foss Field as the balloon ascends. (Courtesy Joel Strasser)

runways gleamed in the bright sunlight. The winter sky was so clear, so vibrantly blue, the silvery glare of the balloon would be visible to the naked eye for the entire ascent.

Nick's first transmission, once airborne, regarded a bet he had made with Raven launch crew chief Dick Keuser. Keuser had insisted that the launch would be so smooth that Nick would never even feel the instant of liftoff. And while the operation had been a great deal smoother than the bumpy Litton launch, Nick had felt a slight jolt.

"Tell Keuser he owes me a steak dinner," he said.

Immediately after launch, Raven pilot Frank Heidelbauer took off in a Cessna 180. The most critical phase of Nick's ascent would be at the tropopause where the balloon would first encounter the stratospheric winds, and it was important for the chase pilot to be positioned below the balloon for quick access to an emergency landing site. Heidelbauer headed east and flew big circles (two or three miles in diameter) around the balloon's constantly changing position.

The rate of ascent was entirely under Ed Yost's control, and Yost wanted to get the balloon to altitude as quickly as the balloon's design would allow. The less time Nick spent in the hostile environs beyond the atmosphere, Yost reasoned, the less time for problems to develop. The rise rate from Joe Foss Field to the tropopause was kept steady at 1,187 feet per minute, and this would be increased to 1,241 feet per minute once the balloon reached the stratosphere.

Except for a moderate chill in his legs at about 12,000 feet that caused him to keep the curtained doors shut, Nick reported generally comfortable temperatures inside his insulated regalia despite the frigid winter weather. For the entire ascent he would offer a running commentary in the form of notes for future flights, mostly suggestions for improvements: rework the camera containers to make them more accessible to the pilot, eliminate the Morse code "marks" from the voice track on the tape recorder, change the method of latching the doors.

"Nick, have you taken any pictures yet?" a voice from the ground inquired.

"No," Nick responded. "I'm going to wait until I get a little higher. But I imagine I should be bringing home some pretty good pictures of this trip."

"Better view than the last flight?"

"You can say that again."

A couple of minutes later, Nick reported that every time he would lift his right arm to check the balloon in the wrist-mounted mirror, he would inadvertently deactivate the reserve parachute opener on his chest pack. He asked Marvin McCall to remind him to check the reserve opener once he reached altitude.

"Roger," came McCall's voice. "Nick, can you slow your breathing just a little?" The ground crew would hear Nick's respiration in his voice microphone for the entire flight and would regularly ask him to adjust his breathing rate.

"Uh, Roger."

At about 18,000 feet, Nick reported that he was continuing to rise "straight up" rather than beginning the eastward drift that Dr. Frisby had predicted.

"Maybe you'll come down in Sioux Falls," McCall suggested.

Nick's immediate reply: "Just keep me away from that damn dump."

At 20,000 feet, having opened and then closed the furniture-pad curtains to ward off an intensifying chill, Nick related another minor oddity. "I keep hearing little bumping noises," he said. "What are they? Probably just the rigging creaking."

Dick Keuser came on line: "Nick, are you twisting at all?"

"Constantly rotating."

"What's the temperature up there?"

"I'm at five below."

"Okay," Keuser said. "Slow your breathing, will you, Nick?"

"Roger. Every time I do a little work I start breathing faster."

By the time the balloon had reached 30,000 feet at 12:38 P.M. and had finally begun its expected drift to the east, the temperature had plunged to 21 degrees below zero.

A few minutes later the conversation with the ground crew resumed. "Has your suit begun to pressurize?" Keuser asked.

"Negative."

"What's your altitude?"

"I'm reading 36,000. And I'm beginning to get a little ventilation."

"Nick, your suit should be pressurizing by now."

"Roger. Yeah, it is."

"Nick, you're eleven and a half miles east of the airport."

"Roger."

By the time the balloon had reached 40,000 feet and the Dave Clark Special was fully pressurized and holding steady, Nick began to allow himself to believe that it was all actually going to work this time. Everything was going off according to plan. As he passed through the tropopause, he felt only a very slight tremor run through the gondola. Conditions were perfect. With a little luck, in another couple of hours, the dream would become reality.

Marvin McCall's calming voice came on line again at about 50,000 feet. "How's it going there, Nick?"

"Real good, Marv," Nick acknowledged. "Hey, I've got a question, buddy. When you get a little perspiration down around the bottom of the mask, how do you get rid of it?" Nick followed this question with a distinct chuckle, which was accompanied by more chuckling from the ground. Clearly this issue had come up before.

Marvin McCall answered, deadpan: "You raise the visor." McCall's laughter made it clear that he wasn't serious.

"Okay," Nick said. "Let me try it."

"Negative," McCall said, sharply. The laughter stopped.

Nick went on to describe the situation. "It's not bad, but just as I lift my head back a little, I get a spray of perspiration over the inside of the visor."

"It's drying off, isn't it?" McCall asked.

"It's like I've got a puddle in here. There's two extremes: warm on the top half and cold on the bottom. There's no way to get this out." After a pause, Nick assured McCall and the rest of the ground crew that he wasn't contemplating opening the visor. "I was only kidding."

At about 55,000 feet Nick reported being able to distinguish the earth's curvature and noted that the sky overhead had become an astonishing, inky black. "It's just ... very, very black. Very black." He was disappointed—as previous stratospheric aeronauts had been—at his inability to distinguish stars against such blackness and made a mental note to work on a method to eliminate reflected light from the huge balloon for future flights.

While warmer than the preceding days, it was now only 6 degrees Fahrenheit on the ground, and the crew that had just landed at the Estherville airport (Ed and Charmian Yost, Janice, and Jacques Istel among them) may have suffered more from the cold than Nick who was reporting minus 12 degrees, but mostly comfortable conditions, aloft. Huddled around the radio gear in an unheated room of the small airport building, they were all bundled up in heavy winter wear—one of the National Guard guys back in Sioux Falls had loaned Janice his insulated flight coverall, which she wore over her New Jersey parka. Yost, Istel, and Darrel Rupp manned the communications gear, while Janice and Charmian took turns tracking the balloon with a pair of binoculars from the best vantage point they could find: a window in the airport bathroom.

With the balloon somewhere around 65,000 feet—now well above the Armstrong Line—and with optimism growing, Janice was beckoned to the microphone.

"God bless you, honey," she said. "I'm real proud of you."

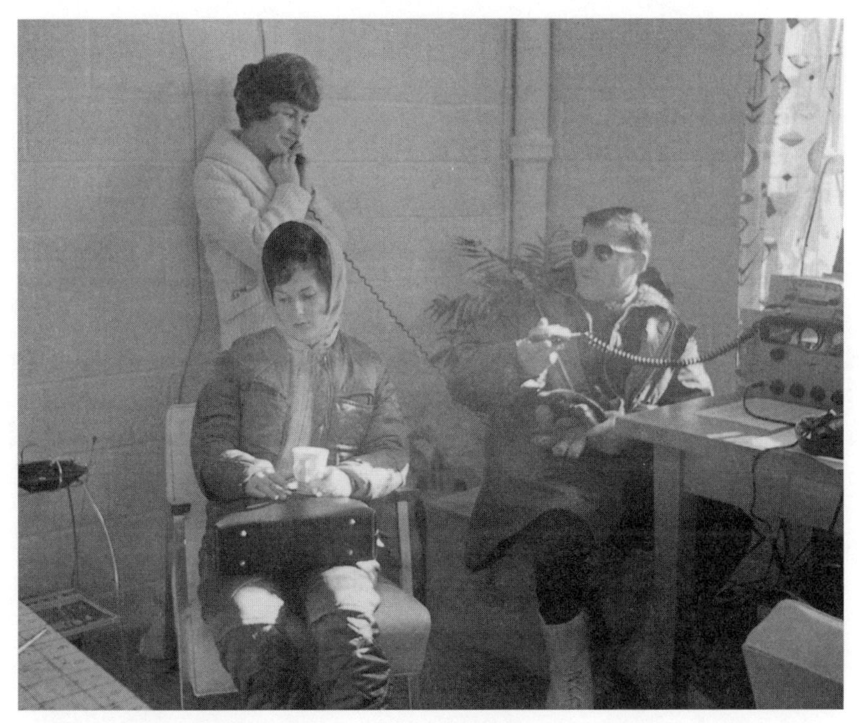

Janice *(seated)*, Charmian Yost, and Jacques Istel in the makeshift communication room at the Estherville airport. (Courtesy TimePix)

"Thank you," Nick answered. "I'll see you in a little while, baby."

Jacques Istel was the next to offer congratulations. "We're all with you, Nick. You're doing great."

"I'm not doing anything, Jacques," Nick said. "I'm just sitting here. The balloon's doing all the work."

"I can't believe it," Istel went on. "I didn't know balloons went up that high."

"I told you. You have to believe me and have faith in these people once in a while."

A few minutes later, with the balloon about thirty miles east of Sioux Falls and tracking nicely toward Estherville, Nick mentioned that the gondola was continuing to rotate steadily. Istel came on line to inquire whether the motion was enough to make Nick dizzy. Nick dismissed the idea and steered the conversation to his panoramic view of the countryside. He had scattered clouds to the north, a bit of haze to the west, but the vistas to the east and south were crystal clear.

"It's just fantastic!"

As he rose higher, through the ozone layer at 70,000 feet and deeper into the stratosphere, Nick observed a phenomenon he referred to as thick "layers of haze." He encountered the first of these layers at about 88,000 feet. As he reached 91,000, he was able to distinguish a separation forming two distinct layers. He noted that the skies in the clear regions between the haze layers were an intense indigo, contrasted with the deep blackness above and below. The black sky now extended down to approximately 10 degrees above the horizon. The visual field beyond the gondola—imagine a bright, burning-yellow sun against a sky blacker than any you've ever seen—was becoming otherworldly and, as had been the case with earlier stratospheric balloonists, served to isolate the pilot psychologically from his crew on the ground.

At 10,000 feet Nick had first armed the automatic opener on his reserve parachute. Now, as he passed through 90,000 feet, he rearmed the reserve and armed the opener on the main Para-Commander canopy. His chute rig was now configured and ready for the big jump. Even if something went wrong and Nick was to lose consciousness at some point during the jump, the parachute system should be capable of lowering him safely to the ground.

Moonrise occurred at 1:30 P.M. for Nick (about an hour and a half before it would become visible to those on Earth), but the moon remained far below, never rising higher than 10 degrees above the horizon while Nick was aloft and could be observed only through the murk of earth atmosphere. The sheer power of the sun he encountered in the stratosphere got more of Nick's attention: "Even with the sun visor down—and the sun visor is very, very good—it was difficult to look to within 25–30 degrees of the sun, and even then it was quite blinding due to the lack of atmosphere. The rays were most intense and heat was felt through both visors. The heat was very, very intense."

Nick had the impression that he was now truly entering outer space. He was nearly three miles higher than the height from which Andreev had jumped and two miles beyond the height from which Peter Dolgov had leapt to his death. He was passing through the altitude Joe Kittinger had attained on *Manhigh I*. And then he passed through the magical 100,000-foot barrier. He was only the fifth human being to make it this high without an engine.

Jacques Istel came on line again: "Nick, you've already got objective one. You've already got a world record. So, if anything happens, you jump right away." Istel had leapt to the conclusion that the free-fall record was already in the bag.

"Yeah," Nick responded, sounding a little irritated, "but we want to go all the way."

Istel continued: "You bet we do, but we've already got objective one. Even I'm getting optimistic."

"You've got to let Ed Yost work on you a little bit."

The balloon continued to rise. And Istel, euphoric that SPACE, Inc. was on the verge of triumph, couldn't quite get over it: "Welcome to the club, Nick. You're a world recordholder!"

But Nick, dangling below a plastic bag full of gas eighteen miles up and with so much yet to do, was not in any mood to celebrate. "We haven't done anything yet, Jacques. Let's wait a little while. Let me know when I get to 110,000 feet. I've got to activate the General Electric oscillator."

Two minutes later and Istel was transmitting again. "Give us a voice test, please."

"Roger. I've got an inside gondola temperature of, oh, 14 degrees. It's warming up."

"Nick," Istel asked, "say a few words to the radio network. We just told the world that you're at 107,000 feet."

"Roger, Jacques," Nick said for the benefit of a local radio audience via telephone hookup with the Estherville airport. "Everything's real fine up here. Inside gondola temperature is now 15 degrees." It was an uncharacteristically tight-lipped public pronouncement from Nick.

"Okay, Nick," Istel said three minutes later. "110,000 feet."

Janice's voice was the next to come on the line: "We're doing great here. How are you doing?"

"Real great, babe. I'm at about 110,500. Somewhere right in that area."

Istel took the microphone again. "Okay, Nick. As of this moment you've got Kittinger."

"Stop it," Nick said, clearly annoyed now. "We don't have anything yet."

"You've got the height!" Istel argued.

"Roger."

"I'm counting on the force of gravity."

"Real good."

"I feel like I've been hearing you breathe for years." The normally combative Istel had gone almost giddy. "I never knew it could sound so good."

Nick, meanwhile, was encountering another layer of haze, this one approximately 2,000 feet thick. He would note yet another layer at 118,000. And still the balloon was rising. The tinny voices of his ground crew, now so far below that they seemed to issue from another world, informed him that he was a full mile beyond the official world manned-balloon altitude record of Ross and Prather on *Strato-Lab V*. He was now very much alone, higher than anyone had ever been in any type of balloon. He spotted Estherville below and even located the little airport. It was a raw, bright winter's day in northern Iowa, bleak and beautiful, and from Nick's perspective civilization was hugging the ground as if it were rooted in the soil like the crops. He identified the three lakes to the east of Estherville, the largest being Okamanpeedan Lake, which straddles the Minnesota-Iowa border. As Nick surveyed the countryside, he found he could pick out individual farmhouses by the thin columns of smoke rising from their chimneys. He could just barely see cars and trucks inching along the roads, but not the General Electric van parked on the side of a highway somewhere southeast of Estherville. Oddly, he was never able to catch sight of any of the small airplanes buzzing about like gnats beneath him.

As Nick approached Estherville, the transmissions from the ground began to cut out due to a phenomenon known to radio operators as a "cone of silence." It wasn't entirely unexpected. A cone-shaped dead spot directly above a radio beacon can, when conditions are right, severely reduce the intensity of all transmitted signals.

"Just tell 'em to let me know when they want me to pull the seat belt off," Nick told Istel.

"Roger . . ." The rest of the response was garbled.

"Say again," Nick barked. "I can't read you. I'm getting intermittent transmissions. I can't understand you." When he got no reply, Nick tried again: "Jacques, test your mike. Try that one again."

Still, Nick couldn't make anything of the choppy response crackling through his headset. He decided to simply report his status and hope that all was well.

"Temperature is good and so is everything else. I thought I heard the word 'altitude.' But I've got to guess because I can't tell exactly where she's sitting. I'd say I'm reading somewhere around 117 or 118,000."

Moments later, Nick had more news to report. "I've got some, uh, powder. It must have popped loose with the balloon expanding. It drifted down and looked like a cloud. There, I'm getting some more. I wish I had access to that camera. That sky is jet black. I'm sitting looking right at it. It's just fantastic."

The response was garbled.

"Go ahead. I can't read you, Jacques."

The follow-up was no more intelligible.

"Everything's okay," Nick announced. "I can't read you, so I'm just keeping up a steady line of chatter so you know everything's okay."

A few minutes more, and the quality of the transmission improved somewhat.

Istel: Can you read me now, Nick?

Nick: Roger, Jacques. Loud and clear.

Istel: Okay, Janice is . . .

Nick: Naw, I got you intermittently again.

Istel: Janice can see your balloon.

Nick: All right, real good. You mean *our* balloon.

Istel: Roger! You are now at 120,000 feet!

Nick: Roger. I'm gettin' vapor. It's coming out of the gondola some place. I can see it filtering out into the outside air.

Istel: Check your pressures and temperatures.

Nick: Temperature is at 24. Oxygen's good. That vapor must be from the cryogenic system.

Istel: All right.

Still the great Raven balloon, which Nick could monitor whenever he liked through the top window by glancing at the circular mirror on his right wrist—he could see that it had swollen now into an almost perfect globe—continued to ascend. Having opened the curtains, swung open the front door panels, and snapped back the front section of the roof, Nick could peer out and down past the toes of his boots at an impossibly wide expanse of midwestern farm country, from this altitude an intricate patchwork quilt of tiny tan- and chocolate-colored mile-square sections dusted in random patterns with sugary snow. The flight's trajectory had squiggled generally east along the state border.

From the top of the sky: 123,500 feet. (The edge of the balloon and the blackness of space beyond are reflected in the mirror on Nick's right wrist.) (Courtesy TimePix)

At 1:53 P.M., *Second Chance* passed through 120,500 feet. About a minute after that, Ed Yost's distinctive voice came on line. "Nick," he said, "we're coming up to a ceiling of about 122,000 now."

"Roger."

Then Istel was back on: "Your breathing is speeding up slightly, Nick."

"Roger, I'm getting a little bit of vapor out of this cryogenic system."

"Try shutting it off," Istel suggested.

"I can't," Nick replied. He found that he couldn't manipulate the needle valve with the pressure gloves. "I can't read it. I can't even get turned to look down at it."

"Okay. If at any moment you feel the urge, just terminate and jump. Anything from now on is a complete success, so don't take any chances."

"Roger," Nick said, thinking ahead now to the jump. "Ed, how're we lookin'? Ed Yost?"

"He stepped out," Istel said.

The plan had been for Nick to remain at float altitude for at least ten minutes before initiating the prejump procedure, but everyone was starting to get perceptibly anxious and Nick made the unilateral decision to move the timetable forward. Of greatest concern to him was the rate of the gondola's continued eastward drift that threatened to take it beyond the range of the General Electric tracking van.

"I'm going to start the countdown," he announced.

This was finally it. The five-minute countdown to the greatest jump of all time. Nick requested that his altitude be radioed to the G.E. personnel who needed a precise float altitude for their jump-velocity calculations. Affixed to the interior of the gondola's frame was an eight-inch-square brass plaque on which was engraved the prejump procedure. At the five-minute mark—per the written instructions—Nick stowed and secured the gondola's cameras and started a voice tape recorder.

More of the garbled crackling came across Nick's headset.

"I hope you're not talking to me, Ed," Nick said, a new urgency in his voice, "because I didn't get it."

More garbled transmission. Radio communication was becoming more of a burden than a benefit.

"I can't read you, Ed. I'm going to keep counting down from Jump Minus Four, okay? I can't read you, but everything's under control."

At J -4 (jump minus four minutes) Nick double-checked the parachute opening activators.

"Going on Jump Minus Three. Visor heat on full. Visor's [Nick refers here to his sunshade, not the main visor] down. Jump Minus Two and a Half. I'm activating G.E. number 2 plug. The plug is activated." According to the established procedure, at this point Nick was supposed to ask for final clearance to jump, but given the severity of the radio problems, he decided to dispense with this bit of protocol.

"Mark," Nick called out. "Jump Minus Two. Releasing the seat belt. Seat belt released." He started a stopwatch and cranked the electrical heating unit on his suit to its highest setting.

"Jump Minus One and a Half. Activating bailout unit . . . if I can get it. There. Bailout unit activated." He took a short break to run a few feet of motion-picture film using the camera trigger switches.

His next task was to disconnect himself from the gondola's oxygen supply and—in the words of his own pilot checklist—"Hit *Life* Magazine camera switch and Go Man Go!" But it wouldn't be that simple.

"Disconnecting ship oxygen . . . uh, having a problem with it."

Disconnecting from the onboard oxygen supply required Nick to separate a small hose running from his pressure-suit controller to an oxygen tank at his feet in the right-forward corner of the gondola. The coupling, Firewel part number F 2250, was an A-N (Army-Navy) approved spring-loaded quick-disconnect fitting. Nick needed simply to pull a sleeve back and separate the male from the female section of the connector. But when he attempted to pull them apart—a tricky operation with the bulky pressure gloves on his hands, but one that he had performed numerous times in decompression chambers and again on Strato-Jump I (albeit, on that flight, with unpressurized gloves)—he discovered that the two sections were stuck. He pulled again. And again. He was able to get a good grip on the collar of the disconnect, but he couldn't get it to slide back. He applied all the pressure he could muster in both directions. He twisted the fitting. "Having a little problem with disconnect when I try to pull it."

Even though they were momentarily unable to transmit to the gondola, those monitoring the voice transmission on the ground could hear quite clearly that Nick's breathing had begun to race. The tone of his voice was changing as well. Gone was the nonchalant aviator's cadence, the military etiquette.

As the seconds ticked by, Nick continued to struggle with the fitting. "We're going a little bit over. I'm having a problem with this disconnect to the oxygen. Ground control, do you read me? We've got problems."

The balloon's rise had slowed to almost nothing as it approached its physical limit and equilibrium. Nick tried everything, but the pressure gloves made any sort of fine manipulation nearly impossible. He banged the fitting against the main oxygen cylinder. He banged it against the instrument panel. He even unhooked the safety strap across the door opening and used the female half of the latch to try to pinch the fitting in order to loosen it. "Isn't this a bitch? . . . Can't disconnect the oxygen. . . . I don't believe it. I can't separate the hose."

On the ground, personnel in both Sioux Falls and Estherville discussed the situation. Jacques Istel's voice cut through the static. "Nick, what is happening?"

But Nick was too preoccupied for conversation. "Oh God, let me get this hose!" Nick's breathing was extremely labored now, closer to panting. "Don't make me talk. I don't believe it. I just don't believe it."

It was now past 2:00, the outside air temperature was 0 degrees Fahrenheit, and the balloon had reached its float altitude, a spectacular unofficial world-record height of 123,500 feet (21.21 miles) directly above Lake Park, Iowa. But no one celebrated. They needed to get Nick out of the gondola, and they had to be extraordinarily careful in doing so. If he pulled a connection loose and damaged the seal of either the pressure suit or the helmet, it would all be over in seconds.

Janice continued to track the balloon with her binoculars from the bathroom window in Estherville. "I could see it," she recalls. "It looked like a fifty-cent piece in the sky." At one point she went back to the communications room to listen to the running dialogue. And even though she'd promised not to intercede, Yost could see her anguish and handed her the microphone. According to Janice: "I think he said, 'Hi, honey. I'm okay.' And I think I said something like, 'I'm praying for you. Love you.' It was very brief, but it was very special to me."

"How in the hell can this thing be stuck? I need a goddamn crescent wrench." Nick was hyperventilating and literally losing his cool. "It's getting so damn hot. . . . God, I just don't believe it."

Not only had Nick turned up the electrical heating on the pressure suit and on the visor in preparation for bailout, he had also disconnected the onboard ventilation system and found now that he was unable to reconnect it. Adding to the temperature problem, the balloon and gondola were no longer rotating at float altitude as they'd expected. The door and curtains were open to accommodate Nick's exit, facing southwest and exposed to full solar heat. In addition, the exertion of Nick's struggle with the oxygen fitting had generated additional body heat that had been trapped within the suit. He was drenched in perspiration.

"Ground Control, do you read me? We've got problems. I can't . . ."

Ed Yost was standing grimly by in Estherville, the flight controls before him. The realization was dawning that he was going to have to cut the balloon away from the gondola and bring Nick down inside his aluminum-styrofoam cage beneath the huge cargo chute. The problem was, of course, that no one had ever attempted anything like this with a human being aboard. There was no way to know what the opening shock of the parachute would do to the gondola, much less the pilot. They weren't even sure whether there was enough air at such an altitude to adequately fill a parachute.

Meanwhile, Nick refused to give up. "I'll try one more time. . . . What a stinking . . . Please make it come loose!"

Yost's steady voice came over Nick's earpiece ordering him to reattach his seat harness as well as the safety belt across the front door opening in preparation for a drop inside the gondola. But there was a problem here as well. With the bulky pressure gloves, Nick was unable to click either the door-belt or the lap-belt latch into place. It was as if, he said, he had his hands inside footballs, which wasn't far from the truth: electrically heated mitts over the inflated pressure-suit gloves, with foam-rubber-lined survival mitts over the electrical ones. He'd never once practiced reattaching the seat- or door-belt. It was a contingency for which the project found itself completely unprepared.

"I don't have the safety . . . I can't hook up the belt. . . . There is no way of making it. . . . Nobody will believe it."

"Try, Nick," Yost barked. "Try!"

"I've got no control with the gloves. . . . Can't do it." His voice revealed his exhaustion. Even a few seconds of sustained effort required a concentrated act of will. By this point, Nick was beginning to accept the inevitable.

"Let's hope this chute opens right away. I just can't release this gadget. Let me know if you're going to cut me down. I can't close the doors, either. I'm very, very immobile. I can't release this. . . . When you cut me down . . . Oh, good night. Why? Why? Why?"

In the Estherville airport, Janice, unable to bear the tension, broke down. A few moments later, Yost cut in again and instructed Nick to wedge his boots into the corners of the gondola and to punch his arms through the styrofoam side pieces and brace himself against the frame members on each side. But with the big padded gloves, this also proved impossible. Nick grabbed two handles that had been attached to the interior frame at either side of the doorway. If *Second Chance* were to buck or tumble either before or during the cargo chute opening, Nick might very well be ejected. And with his oxygen line, literally his umbilical now, still connected to the onboard supply, he would almost certainly rupture the pressure suit and decompress many miles above the Armstrong Line. The in-gondola drop was not the scenario any of them wanted, but at this point the team was out of options. Years later, Raven's Russ Pohl recalled the concern on the ground about what might occur when the big chute opened: "We thought the gondola might just go over and he might get pumped right out of that puppy and pull that oxygen tube right out of his suit. It was pretty tense."

"Nick, I'll give you a countdown from 10. Brace yourself."

At about this time, in continuing observance of Murphy's Law, the radio link with Yost and Estherville began to crackle and break up. After a few minutes of radio silence, the ground crew reestablished communication by switching back to Raven ground control at the Sioux Falls airport. It is hard to say who found those minutes the most excruciating, but a good candidate would be Janice Piantanida, who was struggling to regain her composure.

"Who's going to cut this down?" Nick asked. The balloon had hit float altitude in Iowa airspace, but by this point it had drifted well past Estherville and had crossed back into Minnesota. "Can't hear you, Ed. Let Sioux Falls give me the signal." Then he added, "I don't have any belt on and I can't get one on. I'm starting to really heat up. I have high temp and no ventilation. I can't work the knob with the gloves. But I'm all set."

A voice came on line from Sioux Falls. "Nick," Dick Keuser announced, "we're cutting you down on Yost's signal."

"Okay, Roger," he said. "I can't read Ed at all. I'll just sit tight. When it comes, I'm braced." Nick stuffed his hands into the "stand-up grips" that were positioned to help him raise up prior to the jump. He wedged his left leg against the cryo unit and his right leg against the main oxygen tank.

Although Nick was unable to hear it, Yost's countdown began: "Okay, Nick, mark . . . 10, 9, 8, 7, 6, 5, 4, 3, 2, 1."

At 2:15 P.M., squibs fired, simultaneously collapsing and releasing the great balloon. The gondola dropped like a rock. Figures supplied by the Air Force had suggested that the big cargo parachute attached to the top of the gondola might not deploy immediately, that in fact Nick might fall for tens of thousands of feet, perhaps reaching near-supersonic speeds, before the opening would occur. There would be no external camera in position to record the sequence of events and the behavior of the parachute.

Nick never heard the squibs fire but was suddenly very aware that he was descending and descending rapidly. The initial seconds of the drop were terrifying as Nick and the ground crew waited, poised for the opening shock. Some fifteen seconds after the balloon had been separated, Nick's astonished voice cut through to the ground stations.

"Do you read me anybody? Let me tell you, I'm coming down like a banshee!"

After being separated from the balloon, the gondola in free fall tilted forward at a 45-degree angle as a result of the snap of the balloon release, pointing Nick fully earthward with the doors wide open, but the gondola did not—as feared by some—begin to tumble or spin. In fact, it dropped cleanly, seemingly in verification of the project's belief that a falling body with no aerodynamic force to upset it would fall in the position in which it had been released. Most—including Nick, who counted five separate opening shocks in quick sequence—believed the full parachute opening occurred somewhere around 95,000 feet. The jolt to the gondola, delivered by the opening of the chute, was not the devastating blow—or series of blows—some had feared, but neither was it trivial. In fact, one of the shocks later estimated at around 5Gs was powerful enough, according to one of his ground-crew members, to quite literally knock the crap out of Nick Piantanida. (This little embarrassment, for obvious reasons, would be kept very quiet.)

Nick began to describe his ordeal, not quite in real time, but only moments after the stunning sequence of events. Given his predicament, it's a remarkably composed, measured, and observant performance.

"The parachute dropped right there in front of the gondola, and the gondola started facing forward. The parachute started pulling up. I thought I was going to leave it, but it just rocked back. I'd estimate I was pulling four or five Gs. It [the opening] was about like a Para-Commander. I'm reading about 94,000 now on my absolute pressure gauge. What else? I just want to keep talking so you know I'm coming down. My body temperature is starting to drop as we descend. The gondola temperature has now dropped. It was up to 60 degrees at float; we're down to about 14 degrees above zero. My absolute pressure gauge now reads 86,000. Coming down pretty quick. My oxygen is still good."

Nick's first-hand account of the gradual deployment of the gondola's cargo parachute is fascinating. He would say that the sight of the parachute floating in front of the gondola was his biggest scare of the whole flight. "It kind of shook me up a bit. I thought that this was probably the end of the road. I didn't really know whether or not the parachute would open."

There was not much useful data available in 1966 on the behavior of parachute canopies at those altitudes. But subsequent studies have shown that parachutes in high-altitude environments tend to billow open and then partially collapse before opening again, often several times before the final opening. And even though Nick, who knew well the sensation of free fall, reported being in free fall for a full 30,000 feet through the stratosphere, it is impossible to say with certainty precisely what occurred and in what sequence. Jim Winker's extensive experience with subsequent high-altitude unmanned balloons released even higher than Nick's show that parachutes can in fact open cleanly in a near-vacuum, and Winker believes that the Strato-Jump gondola's parachute probably opened after a brief flare-and-collapse sequence, though Winker adds, "I'm sure Nick *felt* like he was in free fall." Ed Yost is less sure: "No two of those things ever behave the same way. You can't say for sure just what happened up there." Whatever the truth—and Nick's detailed eyewitness account of the incident as narrated during his descent and in his formal project report recorded shortly after the flight must be given the benefit of the doubt—there is little question that due to the thinness of the atmosphere, *Second Chance* became the fastest elevator in history.

Fortunately, the gondola's orientation with respect to the ground remained more or less constant for the early phase of the descent—Nick believed that as the parachute had floated up, one of the risers had briefly snagged a corner of the gondola and righted it. At any rate, Nick was able to keep himself wedged safely inside. Which is not to say that he had a comfortable ride down. The air temperature had dropped rapidly during the descent, and the moisture inside the suit quickly turned an overheated environment into a soggy and near-frigid one. But worse, as the air density increased, the big parachute began to oscillate and whip the gondola violently from side to side and front to back. Pioneer had provided the 46.5-foot chute, a standard military-style cargo canopy that had been manufactured about the time Nick had made his first jump at Lakewood in 1963. And because no one had ever envisioned a pilot descending beneath it, the potential for oscillations had never been seen as an issue.

"I had to hold myself in my seat," Nick would say later. "I'd be looking directly at the ground, then up into the black sky, and at the same time oscillating the other direction, side to side. It wasn't a steady pendulum. I was all over the place." It was as brutal a parachute descent as he had ever experienced.

"You okay, Nick?" Dick Keuser asked.

"It's just a little hairy," was the best Nick could muster.

"Just stay with it."

"I can't get out of it!" Nick said, his fatigue taxing his patience.

"All right," Keuser said, wanting to keep Nick talking. "You still there?"

"I'm still in it. I wish to hell I could get out of it."

"When you get down lower you can get out of it."

Nick was totally disgusted now. "Yeah, right." With his own survival no longer so alarmingly at stake, the full impact of what had transpired at float altitude was setting in.

"Damn!" Nick said to no one in particular. "Damn! Damn! How can one little stinking piece of equipment screw up like that?"

"Next time we'll have better oxygen equipment for you, buddy."

"I don't want to talk about it. A *quick* disconnect! The quick disconnect didn't disconnect! What can I tell you?"

"Can you give us an altitude reading? And how do you feel?"

"Aw, I feel lousy," Nick said. "Physiologically I'm good. I'm coming up on 69,000. But I'm gonna shut up. I'm starting to hyperventilate slightly, so I'm just going to keep quiet."

The shock pad beneath the gondola had been dispensed with for this flight—no one had believed there would be any need for it—and that was becoming a worry for the ground personnel.

"Nick," Keuser asked, "when you get low, are you going to jump? Because the gondola has no shock pad beneath it."

"When I get low," Nick replied, "I'm going to jump. Yes."

Shortly after the parachute opening, radio communication had been reestablished with Estherville. At 65,000 feet, Ed Yost came on line to ask Nick for a status check.

"I'm getting sick," he reported, his voice woozy. "These oscillations. Don't make me talk." The worry now centered on suffocation and the risk of Nick vomiting inside the pressurized helmet. Although he didn't mention it at the time, he was already experiencing dry heaves.

"What's your altitude now, Nick?"

Nick's responses were becoming extremely subdued, as if he were awakening from a deep sleep. "Fifty-five."

"The next one will be good, Nick." It was Jacques Istel's voice. Nick did not respond.

Marvin McCall in Sioux Falls tried next: "Nick, this is Marv. What's your altitude?"

"Fifty-two," was the reply.

"How's it going," McCall asked. They were all trying to figure out if Nick was competent to attempt a parachute jump in his present condition.

"Oh," Nick said gloomily, "fine, Marv."

Istel tried once more to engage Nick in conversation, again without success.

"Nick, this is Keuser. How're you feeling?"

"Aw," Nick said, "I'm gettin' seasick with these oscillations."

"What's your altitude now?"

"Uh, thirty-three."

Once the Dave Clark Special deflated at just above 30,000 feet, the gloves became more manageable and Nick began to fumble with the seat belt. He tried again to disconnect the oxygen line, but it remained firmly jammed.

Jacques Istel made another attempt: "Okay, Nick. How do you feel now?"

Nick: "Woozy."

Istel: "Nick?"

Nick: "Mmm?"

Istel: "Listen, take a look at your altimeter. How high are you now?"

Nick: "Uh, twenty-seven."

Istel: "When you get down to about 18,000, Nick, you can jump. Just relax for a few more moments until you're down there. You're almost down there now."

Nick: "Mmm hmm."

The new worry was that Nick would pass out from exhaustion. The consensus was that it was important to keep him talking.

"Okay, Nick. This is Marv."

"Yeah, Marv." Nick's words were almost slurred.

"You're sounding pretty weak."

"Yeah, I'm getting a little woozy. I'm okay, though."

"Is it getting any worse?" McCall asked.

"Naw."

"Do you have complete control of yourself?"

It was a question Nick Piantanida wasn't used to hearing. "Uh, yeah. I hope so." There was a pause as he reflected. "Yeah. Whaddya need?"

It was Istel again: "Nick, you know, if you have any doubts at all, just stay in the gondola. It'll bring you down all right. You'll be a bit seasick, but you'll be fine." When there was no answer, Istel prompted one. "Acknowledge, please."

"I heard you."

"A couple of more minutes, Nick," Istel, the most anxious figure in the whole drama, rambled on, "and you won't even need oxygen. A couple more minutes and you can open up everything. And if you're feeling good, you can jump. If you're not feeling good, just stay in the gondola."

"Roger," Nick replied. He was allowed about a minute of radio silence before Istel, who was desperate to keep him talking, was back on line.

"Nick, are you going to jump?"

"I don't know," Nick said. "I haven't made my mind up yet."

"Well, let me know when you're down to 18,000 feet."

As the gondola's descent continued and the atmosphere thickened, Nick's ordeal gradually became more tolerable. By 16,000 feet, the side-to-side oscillations of the gondola had dampened somewhat.

"It's starting to stabilize now," Nick announced. "But the way it was going before . . . God."

Intense discussions were underway both in Estherville and Sioux Falls about the wisdom of allowing Nick to remain in the gondola versus encouraging him to cover the last few thousand feet beneath the familiar Para-Commander.

"We're worried about damage to your thighs," Istel explained to Nick. It was hard to predict the severity of the gondola's impact with the ground for a man inside seated on what was essentially a plywood box. Eight years earlier, balloon builder Otto Winzen and an Air Force captain named Grover Schock, fearing that winds were about to blow them out over Lake Superior, had cut loose from their gas balloon at an altitude of about 100 feet. The impact had snapped the bones in Schock's legs like toothpicks. "We want you to stand up when you come in," Istel said. "Here's Ed Yost."

"When you come in," Yost instructed Nick, his voice clear and his instructions crisp, "stand up and grab the tubing on top. Brace your feet sideways and push with your feet."

But Nick didn't share all the concern. "Mmm hmm. It's no big deal." Having survived to this point, the prospect of a hard landing didn't seem to him like much cause for alarm.

Yost continued: "Well, it may not make any difference. You're coming in plenty slow. What are you showing now on your [negative] rate of climb?"

"Mmm, it looks like about 800, 900 [feet per minute]."

"You probably ought to ride 'er in," Yost suggested. "If you try to stand up and you feel a little weak, just stay in the seat."

Nick's voice was feeble. "Right."

Yost: "Is your faceplate open?"

Nick: "No."

Yost: "Okay. Are you going to open it at a lower level?"

Nick: "Yes sir."

Yost: "All right. We've got airplanes right underneath you. There's no problem with a landing."

Nick: "My main canopy is going to bust open at about 7,000 feet."

Yost: "Is that the front one?"

Nick: "No, the rear one."

Yost: "Okay. You didn't deactivate it?"

Nick: "No."

Yost: "Well, just sit there and hold her back. That's all you can do."

Nick: "Aw, I think I might go out here when I get down to about 10,000." With the gondola stabilized, Nick's spirits seemed to be picking up just a bit.

Jacques Istel, who was more familiar than Yost with the parachute hardware, took over the Estherville microphone again. "It's [the main canopy] in a sleeve, is it not?"

"Roger," Nick said.

"Then it doesn't matter if it opens in the gondola."

"Uh, yeah. OK."

"When it comes out, sit on it. It'll give you additional cushioning."

"Roger."

"Okay, Nick," Istel asked, "what altitude are you at now?"

"Uh, 12,500."

"Well, if you jump, that's a 60-second delay." After a brief discussion with Yost, Istel decided to try to discourage Nick from leaving the gondola. "We all think it's best for you just to take it easy and stay in the gondola and have an unusual landing."

Yost came on again. "Yeah, this'll be a Piccard-type landing. In the gondola." This was a reference to a series of launch mishaps and premature landings that had befallen Don Piccard in previous years.

"Mmm."

"If you feel okay," Yost reiterated, "you can grab the thing and hang on, right?"

"Roger."

"You sound tired," Yost said. "I don't know if you're mad or tired."

Nick was clearly both. "Aw, I'm just . . . I'm a little woozy."

"Yeah, okay," Yost said. "I think you can get the faceplate open here in just a while. Is the strap across the front still connected?"

"No," Nick said. He even chuckled slightly at his own pathetic predicament. "That's . . . I couldn't get it connected before. I'm going to hook 'em all up now."

"Okay," Yost said, concerned that an exhausted Nick not be tossed around too badly upon impact. "I think that's a good idea. The front strap's less important than the safety belt."

"Roger."

"Okay. We have a very light wind on the ground. It's mostly calm. We have a little wind, maybe five miles per hour, from the northwest."

"I just hope I don't hit one of those houses," Nick said.

"Don't worry about it," Yost told him. "You won't get hurt inside that box. Just hang on."

"Roger." After a few moments, Nick announced that his main chute had popped open behind him.

"Nick," Istel said, "stuff the chute under you. All right?"

"Roger."

At 2:45 P.M., thirty-two long minutes after being separated from the balloon, Nick smacked down hard on the edge of a desolate, snow-covered Iowa stubble-field some thirty-five miles due east of Estherville and equidistant between the towns of Elmore, Minnesota, and Lakota, Iowa, the big red-orange and white striped cargo chute draping itself across a barbed-wire fence. In spite of a wrenched left hand, a right-shoulder muscle pull, and a cut on the chin caused by the radio microphone on impact, Nick landed—in a half-standing position as Yost had suggested—in relatively good shape. Back on the ground, the first thing he did was to dig out a pocketknife and begin to pry furiously at the oxygen fitting that had spoiled his world-record jump.

When he heard pilots on the radio saying that they could see no movement in the gondola, Nick released his seat belt and crawled partially outside to wave at the chase crews so they would know he was all right. Then he reentered the gondola to resume work on the fitting. The disconnect had iced up on the freezing descent, but even after the ice had been chipped away, it took Nick nearly three minutes of banging on it with the knife handle to loosen and separate the two halves of the fitting.

Meanwhile, in a bizarre footnote to the landmark flight's story, a local farmer who had stood in rapt astonishment as he watched the curious silent aircraft with its space-suited humanoid appear out of nowhere, staggered and dropped to the dirt with a heart attack.

As Dr. Leonard Thompson, Project Strato-Jump's flight surgeon, landed by airplane and made his way to the gondola, Nick pointed with his knife to the ailing farmer, directing the medical resources where they were most needed.

"If only I'd had a damn dollar and twenty-five cent wrench," he muttered.

"Never Down for Long"

Janice had waited patiently in the cold room at the Estherville airport for them to bring Nick in. Some of the crew were starting to arrive and she

Nick addresses reporters prior to the press conference in Estherville. (Courtesy *Wichita Eagle*)

overheard someone say that Nick was "really pissed." When Nick finally arrived, haggard and dejected, still in the suit (but without the helmet), his first utterance was, "Where's Janice?" The two of them sat alone in the little room long enough for Nick to smoke a cigarette and drink a cup of coffee.

"Well, honey," he said, "now I know I can get up there. Now it's just a matter of getting out of the damn thing and free falling." To Janice, it sounded like the hard work had all been done and the road ahead was free and clear. She knew without having to hear it that Nick wasn't finished.

"He was never down for very long," she says. "Or maybe he was just putting up a good front for the crew and the press. I know he was terribly disappointed."

Then he was whisked away to remove the pressure suit and get cleaned up in preparation for the first of two press conferences: one at the airport, and one later in Sioux Falls. Reporters were already milling impatiently about the airport, wanting answers to the two big questions: What went wrong? And, are you going to do it again?

Whenever the editors at *Life* magazine came across an offbeat or exotic story—when, for instance, they heard about an eccentric couple who wanted to swim the Bering Strait and get married under water halfway across—they looked for Roger Vaughan. The wiry reporter had a keen curiosity and had proven himself particularly adept with outdoor adventure stories. So when word began to get around about a crazy parachutist from New Jersey who wanted to break the world free-fall record, Vaughan naturally got the assignment.

Vaughan first met Nick Piantanida in New York at Jacques Istel's office and had been bowled over. Nick, with his military haircut and massive shoulders, had awed Vaughan with his physical presence. But it was the Piantanida charm that won the reporter over.

"I didn't know what to expect—maybe some sort of screwball. But here was this big moose of a guy, very calm, with steady eyes. . . . And he had a nice way about him. I mean, he was a truck driver from New Jersey, but he also had a lot of charm when he wanted to dish it out. He could be diplomatic, but he could also be rough-and-tumble. It depended on the situation. He was a well-rounded guy." Vaughan and Piantanida quickly became casual friends, and it wasn't long before Vaughan had been persuaded to make his first parachute jump—under Nick's guidance.

"You had to get up at three in the morning," Vaughan said, "because Nick always left at four. Soon I had the bug. . . . It was marvelous." Later, following the training flights with Tracy Barnes and the flight of Strato-Jump I, Vaughan accompanied Nick and Janice on a series of purely recreational hot-air flights in the South Dakota Badlands near Mount Rushmore.

Both Vaughan and *Life* photographer Robert Kelley had arrived in Sioux Falls several days prior to the *Second Chance* launch and both were present at the subdued postflight press conference. Kelley, a Chicagoan, was an interesting guy and a crack action photographer. But Strato-Jump posed some unusual challenges. To get good shots of the gondola in the air, and—he hoped—of Nick in the final stages of his epic jump, he had rented a plane and hired his own pilot. But if he wanted images from Piantanida's point of view, Kelley realized, he would need a remote-control set-up that Nick could trigger with his pressure gloves. It would also be necessary to ensure that any onboard cameras could withstand temperatures as low as 65 degrees below zero. He'd covered one camera with heater tape after removing all the lubricant—which had shown a tendency to freeze at stratospheric temperatures—and mounted it at the top of the gondola. That one captured some breathtaking images encompassing the vast checkerboard plains from 123,500 feet.

The press conference in Estherville was a brief one. Nick told Roger Vaughan and the other reporters that his oxygen line had jammed and that all he had needed was a wrench. He promised them, with a tight grin that was only half-convincing, that he would indeed be back in the stratosphere in a matter of weeks. The third time would be the charm, he promised: "I'll get up there one way or the other, sooner or later."

By the time the whole entourage arrived back at the Raven plant, a driving snow that complemented Nick's mood had begun to pummel Sioux Falls. Nick met the press again and dourly read them some of the postflight statistics from a sheet Ed Yost had handed him. The flight's duration had been two hours and thirty-four minutes, the total lateral distance covered had been 140 miles. The coldest temperature encountered had been 61 degrees Fahrenheit at 40,000 feet. The peak altitude, as reported by beacon transmitter, was 123,500 feet, and Nick had remained at that height for 20.8 minutes. He repeated the line about the $1.25 wrench. They'd been so tantalizingly close.

"I considered cutting the oxygen hose," he said, warming a bit to his subject and to his audience, "but regarded that too dangerous. I could have clamped the hose in my hand and jumped, but if the hose had got away from me in the fall I wouldn't have lasted more than six seconds. I had visions of being jerked out of the gondola. But this wasn't the day the man up there wanted me."

Even though it would forever be unofficial, the team could claim an absolute lighter-than-air altitude record: the highest manned balloon flight ever made. A Raven press release would suggest that Strato-Jump had captured yet another record: the highest opening-without-delay parachute descent. They argued that even though Nick had remained inside the gondola, he had in fact fallen beneath a parachute. Nick would later ask Jacques Istel to contact both the FAI and NAA about the open-canopy descent record. And while Eugene Andreev's record would remain on the books, in a nice bit of irony, the open-canopy descent record they were claiming to have surpassed belonged to Andreev's companion on the *Volga*, the ill-fated Col. Peter Dolgov.

Nick and Istel also insisted that the flight had disproved two theories about high-altitude parachuting that enjoyed currency with the U.S. military. First, they argued that Strato-Jump II had discredited the notion that a near-supersonic parachute opening was unsurvivable. They'd just done it. *Second Chance* had fallen 25,000 feet and was traveling at a speed of between 550 and 600 miles per hour (subsonic, but still faster than a .45-caliber bullet shot from a gun) when the cargo chute had opened. Second, they claimed to have disproved the belief that an object falling from stratospheric heights without a stabilization device such as a drogue chute will tumble and spin out of control. The gondola had fallen from more than 123,000 feet in a generally stable attitude and had shown no tendency to tumble. The first claim had more merit. The second was suspect in that the Air Force's well-documented conviction that a human being untrained in skydiving techniques had a tendency to spin in unstabilized free fall did not necessarily apply to a falling cube.

Most of the crew celebrated the altitude record, although it is unlikely that any of them could have imagined that the achievement would remain unchallenged into the next century. One of the celebrants was certainly Bob Kelley; he had captured some spectacular images from high in the stratosphere, some of the highest and best color photographs ever taken at that time. There was a lot of postflight congratulation and conversation about the positive aspects of Strato-Jump II.

"What was it like up there?" someone asked.

"It was the most awesome sight," Nick said, "to be up there and look out and see that black sky. It was exhilarating."

But in spite of public pronouncements to the contrary, according to Janice, Nick was not really very impressed with the unofficial altitude record. "I couldn't care less," he told her privately. "We didn't accomplish our mission."

"It *was* an exciting moment for him," she believes, "but it was just that he didn't get to do what he went up there to do."

It was maddening that something as simple as a cheap disconnect fitting could have neutralized the years of preparation and testing. But Nick promised the world that he was not through yet. There would be, he insisted without equivocation, a third attempt. In separate discussions, Jacques Istel was saying that a repeat flight was merely "probable," his enthusiasm for the project beginning to wane. Istel's health, which had been poor since the beginning days of his association with Piantanida, was clearly wearing him down.

After it was all over, Nick and Janice headed back to the Town House for a private dinner, though the champagne and steaks ordered the day before had been intended as a celebration feast. Only one of the final three items on Nick's personal "to do" list following the flight remained available to him now:

- Notify VIP's of new record
- Notify PCA and FAI of new record
- GET BOMBED.

5 STRATO-JUMP III
JADODIDE

Nick vowed that on the next attempt he would go all the way to 140,000 feet, which must have been news to Jim Winker and the Raven engineers who had already calculated the ceiling for the balloon they were planning to build for the follow-up, and that was nowhere near 140,000 feet. But the press coverage stayed positive as Nick's public enthusiasm, regained the morning following the second flight, continued to seduce the region's reporters.

A *Des Moines Register* story began: "Parachutist Nicholas J. Piantanida's 'magnificent obsession' to be the first person to free fall through the atmosphere faster than the speed of sound resulted in a 'magnificent failure' here Wednesday afternoon." Some of the most effusive coverage came in a boosterish editorial in the *Sioux Falls Argus-Leader* which read: "And there it was shimmering in the bright sunlight just above the Sioux Falls airport—an elongated tear-shaped balloon. It was on its way. In the tiny gondola beneath was a young man—Nick Piantanida who had come here from New Jersey to make a bold descent from a high altitude." The writer went generously on to compare Nick to Leif Ericson and Christopher Columbus before concluding, "There is a satisfaction in the realization that here in Sioux Falls for a few fleeting moments we were the center of one of these grand adventures. And, if Piantanida has his way, we'll be in the center of it again before long." Most of the major papers in the region had sent reporters to cover the story, and after the heartbreaking failure aboard Strato-Jump II, Nick received personal letters from a number of them volunteering ideas for publicity. Mostly they wrote encouraging him to try again. Almost no one had gotten enough of Project Strato-Jump.

Meanwhile, the skyward attentions of most of the rest of America remained focused on NASA and the accelerating space race, which rendered the travails of even the most intrepid of balloonists pedestrian in comparison. In June of the previous year, James McDivitt and Edward White had piloted *Gemini 4* through sixty-two orbits of the earth, the first NASA flight to rival the duration of the celebrated Soviet space missions. Of equal significance, White did what Piantanida had been unable to do and left his vehicle for a twenty-one-minute tethered space walk, an American first. The follow-up flight, *Gemini 5*, logged 120 orbits. December saw the launch of two more NASA missions. A little more than a month before the launch of the second Strato-Jump flight, *Geminis 6* and 7 had maneuvered to within twelve inches of each other and had flown in perfect formation for nearly eight hours at a height of 185 miles. Frank Borman and James Lovell went on to rack up a total of 206 orbits.

Then, one day after the flight of Strato-Jump II, at 11:45 in the morning Sioux Falls time, the Soviet Union stunned the world by landing an unmanned spacecraft named *Luna 9* on the surface of the moon. The race to deposit a man on that surface was starting to get interesting.

Taking Stock

The three Piantanida daughters waited back in New Jersey, and Janice was anxious to get home. Most of the Strato-Jump crew packed up to return to their lives, their jobs, and their families. Nick, however, proceeded—only two days after the flight of Strato-Jump II, perhaps the single most disappointing event of his life—to make the first test-jump of Raven's experimental plastic-film parachute. Even though the company had initially been leery of exposing their parachute technologies to Nick because of his close association with Pioneer, they had become enamored of the idea of having a celebrated skydiver put one of its prototypes to the test. The chutes used a canopy constructed of a polypropylene fiber scrim sandwiched between layers of polyethylene and were originally intended for single-shot use for cargo loads. Nick's test-jump proved that such chutes could be used safely by human beings. He reported his descent as "extremely stable." He was able to rotate the canopy successfully by manipulating the risers and found that he could complete a 360-degree controlled rotation in about ten seconds. And while the disposable plastic chute was ultimately an innovation that the skydivers of the world mostly chose to ignore, the jump itself was a rousing success that generated quite a buzz in the halls of Raven Industries. Some of the crew involved in the test-jump remarked that Nick's good spirits seemed to have returned full force.

In the week following the second Strato-Jump flight, another round of thank-you letters went out to the crew and the growing ranks of supporters around the country and around the world. Among Nick's well-wishers were the Presentation Sisters at the McKennan Hospital in Sioux Falls; ex-servicemen from all branches of the military; Teamster brothers who had finally gotten with the program; New Jersey citizens proud of their native son; an admirer in Johannesburg, South Africa; and grade-school and junior-high classes all over America. Requests for autographs and photos were now arriving almost daily, along with prayers and four-leaf clovers and little morsels of chocolate from kind souls everywhere. Nick even heard from a parachutist in Seattle who wanted to make his own superstratospheric free fall and was in need of some advice; it seems the would-be strato-jumper had hit a stone wall in his bid to obtain a pressure suit.

Ed Yost and Russ Pohl of Raven joined Nick in contacting key individuals and organizations that had provided assistance. The general tone of the postflight letters, telegrams, and phone calls was surprisingly upbeat. A note, for example—signed by both Nick and Yost—to the staff at the Estherville, Iowa, airport which had assisted with the establishment of the downrange communications post catalogued the successes of Strato-Jump II: "The performance of the balloon carrier and its emergency cut-down system and parachute was excellent. The success of this part of the program provides a firm foundation on which further flights can be based. We know it can be done!" (Strato-Jump and Raven Industries had become a "we.") The note suggested that the problems encountered with the oxygen quick-disconnect only underscored the potential value to the American space program of field-testing emergency escape procedures and hardware at superstratospheric altitudes: "Clearly it is better to face these problems in a carefully controlled experimental program," the letter argued, "than to proceed without such background experience."

Even though the initial postflight consensus regarding the quick-disconnect had been that the oxygen-line fitting had literally frozen tight in the extreme cold of the stratosphere, subsequent investigations had changed that thinking. Nick's own report documents the project's conclusions: "Abrasion marks were found on both the male and female portions of the disconnect. It was therefore reasoned that some foreign matter had worked its way into the fitting and prevented it from separating. It had to be pryed apart following impact. This took some three to four minutes."

Much as he had done in Minneapolis following the Litton flight, Nick continued to show public support for all his contractors, including the Firewel Company who had supplied and installed the piece of equipment that had doomed the attempt at a world-record free fall. He did tell reporters following the flight that the problem with the stuck coupling was "unheard of," but that was the extent of his commentary on the matter. Perhaps surprisingly, there are no indications that Nick was even privately unhappy with Firewel. Should he have been?

If the postflight analysis that showed scarring inside the connector did in fact indicate that a piece of foreign matter—a speck of gravel, a splinter of shrapnel, or something else—was the cause of the friction that prevented the sections from sliding apart, then it is necessary to at least consider the possibility that negligence may have played a role. A treatise on aviation oxygen systems and their maintenance published in a Lockheed field service digest in 1963 addresses contamination of such equipment with a list of nine cautionary practices, the first of which reads (italics in the original): "*Be certain* that all fittings, valves, regulators, and related equipment are individually packaged in containers sealed to exclude dirt and dust." In fact, Nick later confided to Earl Clifford of Firewel that a piece of brown paper had been found lodged between the two halves of the disconnect.

Is it possible that carelessness in the handling of the hardware contributed to a malfunctioning disconnect? For whatever reasons, this possibility does not appear to have been seriously considered by the Strato-Jump team. If the issue did come up, it must have been dismissed rather quickly. There was apparently never any question about whether Firewel would continue in its role as oxygen-system contractor for a follow-up flight. In fact, the company appears to have begun working on a more reliable disconnect arrangement almost immediately.

Postflight analyses were also conducted and published by some of Strato-Jump's other contractors. Three weeks after the flight, General Electric submitted a summary report. Even though the G.E. team's expertise—measuring and recording free-fall velocity—wasn't brought into play due to Nick's inability to extricate himself from the gondola, its analysis is interesting. While much of the report focuses on the positioning of the company's instrumentation van, some general conclusions about the launch and flight operations are included: "Rate of climb from launch to float altitude was 1230 feet/min. Planned rate of climb was 900-1000 feet/min. This would indicate that an excess amount of helium was valved into the balloon prior to launch. Since that type of balloon had no valve for venting off excess gas, rate of climb could not be reduced." This recalls Nick's own charge of overinflation on the Minnesota flight, but it contradicts Yost's stated desire to get the balloon—which, unlike Litton's, performed perfectly—to peak altitude as quickly as possible.

General Electric's report also reveals that even if Nick had been able to get out of the gondola and fall to Earth at the end of the twenty-minute delay at float altitude as he struggled with the disconnect, no meaningful free-fall data could have been recorded due to insufficient remaining battery power in the doppler transmitter. The report recommends a redesign of the transmitter to reduce power consumption. In a follow-up report to Strato-Jump a month later, G.E. also announced a recalculation of terminal velocity for a fully laden skydiver falling from above 120,000 feet. The new numbers—in conflict with earlier calculations provided by Pioneer Parachutes—suggested that supersonic speeds might not, in fact, be achievable. To reach the speed of sound, Nick would need to fall in a "head-first to airstream" configuration, which would be difficult to achieve without some sort of stabilization device, which in turn would create drag and diminish the descent speed. Falling perpendicular to the airstream, either face-to-earth or back-to-earth, the maximum speed achievable, according to G.E.'s revised calculations, was Mach 1.03. This obviously left only the thinnest margin of error for either the mathematics or for human performance.

Bill Jolly of Pioneer Parachutes had come to Sioux Falls for the launch of Strato-Jump II to oversee the special parachute equipment his company had provided. Nick knew Jolly from East Coast parachuting circles and had told Pioneer after the first flight that of all the expert riggers in the country Jolly was his first choice. He assisted Marvin McCall and others with Nick's suit-up and general equipment preparations. His postflight report contains no unique observations or conclusions, generally agreeing with the analyses of Raven and General Electric. His recommendations focused on modifications to the parachuting rig: tightening the drogue portion of the backpack, adding more adjustable fittings to the harness, reworking all ripcord and timer handles for parallel and consistent operation, and adding a timer and radio unit inside—rather than behind—the reserve pack. Jolly's report predicted that Pioneer could have the modified rig ready to go by May.

Civilians in Balloons

It is useful at this point to put Nick Piantanida and SPACE, Inc. into some historical perspective in terms of funding. For the nearly two centuries of aeronautics that preceded Strato-Jump, sponsorship had always been a thorny issue. Balloon operations were cheaper than other forms of high-altitude aviation, but they were still expensive in real terms. Not only did stratospheric aeronauts have to acquire specialized hardware such as the behemoth plastic balloons and gondolas and survival gear, they had to secure massive quantities of costly lifting gases. Nonflammable helium has been the balloonist's choice for high-altitude operations since the flight of the National Geographic Society's *Explorer II* in 1935. While the lion's share of the world's helium production comes from natural gas deposits in the American Midwest, Southwest, and Mountain West, helium is relatively rare and, consequently, expensive—particularly when purchased in tanker-truck volumes. Modern high-altitude aeronauts also needed special expertise in the form of launch services, including communications and tracking, and ground- and air-recovery vehicles. In addition, if they were wise and could afford it, they did lots and lots of preflight testing.

Some balloonists got their funds (and ultimately their agendas) from agencies of their national governments. Others found private funding and thereby preserved more of their autonomy. During the great balloon-altitude competition of the 1930s, the Soviet flights were naturally government-sponsored while the United States' entries were bankrolled by large private enterprises with only the occasional participation of state agencies. But by the 1950s the game had changed. The cost of raising a manned balloon into a space-equivalent environment had gotten progressively more—even prohibitively—costly, which explains why America's postwar stratospheric balloon projects were military in nature.

Even though the private sector contributed to a number of military programs and, in the case of Project Stargazer (Joe Kittinger's astronomy-by-balloon project of 1962), participated as part of a funding coalition that included the Air Force, major universities, and several corporations, all postwar stratospheric manned balloon projects had been government sponsored—with a single exception. Nick Piantanida and Jacques Istel had planned, funded, tested, staffed, and completed two flights—one achieving an unofficial world-altitude record—and they had basically done it themselves. It was the closest thing to a civilian space program the world had ever seen.

Traveling horizontally would always be less onerous. Private funds could still back balloon distance- and duration-record quests. If a man wanted to attempt a crossing of the Atlantic Ocean, for example, it was quite possible for him to fund it himself. The problem with private funding is that it can facilitate ill-conceived operations that have only a limited—if any—chance of success. Raven Industries would become involved with one such flight.

In February of 1974, eight years to the month after the flight of Strato-Jump II, the self-funded Thomas L. Gatch launched from Harrisburg, Pennsylvania, on an attempt to complete the first balloon crossing of the Atlantic. Gatch's unusual transportation system consisted of a cluster of ten 5-mil polyester superpressure balloons, each twenty-six feet in diameter, and a spherical fiberglass gondola outfitted with polycarbonate window ports.

But Gatch had either been sloppy in his calculations or lax in self-discipline and had launched with a payload weighing substantially more than his plans specified. In spite of the fact that he had been supplied with precise weight calculations and recommendations by Raven Industries, which built the balloons to his own specifications, Gatch took off with a payload weighing some seventy-six pounds over the maximum allowable. Somewhere above 36,000 feet, one of the balloons burst, probably due either to overinflation or to chafing against adjacent balloons.

With only nine balloons left, Gatch would have needed to jettison 157 pounds of his total payload in order to maintain his altitude, but he had provisioned himself with a mere ten pounds of disposable ballast. Gatch had consulted regularly with former Litton Industries' balloon manager Karl Stefan and had bragged about how little his system had cost him. When Gatch's last known radio transmission was received, he reported his altitude as 36,000 feet and declared himself 1,850 miles from Harrisburg. He was sighted by a ship on the third morning of his flight, but that was the last anyone saw or heard of Tom Gatch. Like most of the dozen or so transatlantic hopefuls who had preceded him, he disappeared without a trace.

Had Gatch been associated with a state-sponsored project, there's little question that many of his problems attributable to a lack of experience, knowledge, and operational discipline could—and likely would—have been avoided.

Doubt and Resolve

Nick Piantanida confided to his wife that while he felt compelled to continue with Strato-Jump out of allegiance to his crew and to all the benefactors and supporters and friends who had sacrificed for the project, he no longer felt as personally driven to make a third flight. The ground seemed to be shifting out from under Strato-Jump and just about everybody connected with the project could feel it. The American cultural landscape was on the verge of a major overhaul—new political assassinations, big city race riots, massive war protests—and Nick's dream had somehow lost part of its significance, its sense of mission. He had gotten plenty of positive attention with Strato-Jump and had tasted a few of the seductions of celebrityhood. He and Janice had talked privately about Hollywood. Nick had the looks, the charisma. One of his financial backers had observed that he resembled and carried himself like a white Muhammad Ali. Maybe they could use him in the movies. Nick also talked of learning to fly sailplanes, having already joined the Soaring Society of America, and about his desire to become an astronaut—specifically, to land on the moon. He understood that his 6-foot-1-inch, 210-pound frame was too large for NASA's first-generation space vehicles which was why he'd already begun working on a design for a one-man, pilot-controllable rocket that could accommodate his bulk along with his ambitions.

Still, he had promised the public a third Strato-Jump flight and he felt honor-bound to make good on his word, he told Janice. He promised her that this would be the last time. Janice did what she had carefully refrained from doing until now and asked him, pleaded with him, not to go forward with Strato-Jump III.

"By that time we had three children," she explained. "And the reality sank into me that there was a possibility of things going wrong. I had been so scared the other two times." Janice had continued to be the focus of a sustained lobbying effort from the Piantanida family. "His mother had really gotten to me. She wanted this stopped. And I think Vern had gotten to me by that time as well. I was the only person who maybe could have talked him out of it. Nick wouldn't listen to anybody else. And I remember having a lot of time with him before that last flight. He was more preoccupied than before. And I remember thinking: 'I wonder if he thinks this might not work. That he might lose his life this time.'" An element of Nick's private apprehension about a third flight must surely have been the result of the vulnerability he had experienced on his previous flights and the knowledge that there was a very real possibility of the plans and checklists breaking down in the face of equipment failure or other problems beyond his control.

There was at least one other openly dissenting voice. Jacques Istel had been trying since mid-February to convince Nick to quit while he was ahead. This stratospheric business was not only too dangerous, Istel maintained, it was too expensive. Raven pilot Frank Heidelbauer had invited Nick and Janice to his house on the east side of Sioux Falls for dinner that spring, and he recalls a phone call during which Nick argued loudly with Istel.

"He was always trying to talk Nick out of this venture," according to Heidelbauer. "They'd have some fierce arguments. But Nick would say that he'd raised money to do this and by God they were gonna do it!"

Few connected with Project Strato-Jump, including Nick, knew the true extent of Istel's medical problems. Due to his aloofness and a general reluctance to mix with the rank-and-file, Istel had remained mostly a mystery to the troops in Sioux Falls. Many of the participants remember him as sickly, and only the insiders were aware that he was almost always accompanied by a private nurse who monitored his condition. What practically none of them knew was that the great parachutist had suffered from a severe and painful pancreatic disorder since the early 1960s that ultimately required two major operations. Frequent injections of the narcotic Demerol were required to counteract the constant pain. It is reasonable to assume that these health difficulties, along with the Demerol, colored not only the perceptions of the Strato-Jump and Raven personnel of Jacques Istel, but that they also affected Istel's desire to continue with what had become an increasingly pressure-packed and mentally draining effort to achieve an extremely elusive goal.

But projects like these have a way of taking on a momentum and life of their own, and whether Nick was totally committed or not, whether Janice and the rest of his family were willing to put themselves through the emotional ringer again or not, whether Jacques Istel approved or even continued to be involved or not, Nick knew he was going to have to make another attempt on the Soviet record. On Joe Kittinger's record. "Three," after all, had been the magic number for Project Excelsior. Not until his third attempt had Kittinger made the leap that had landed him in the record books. Perhaps "three" would prove lucky once again. Nick, the father of three daughters, was thirty-three years old. But Project Strato-Jump would need much more than a lucky number. It hadn't, after all, done America's first stratospheric aeronaut any good at all: Hawthorne Gray's third flight had killed him.

In an interview with an Associated Press reporter who had tracked him down in Brick Town that spring, Nick gave the impression of being supremely unconcerned about his own fate and had laughed openly about how he had "cheated death" on his previous flights. He insisted that he was still committed to his dream, that he continued to work seven days a week on the project, with only four hours a day set aside for sleep. He revealed that he had recently added to his daily ritual a regimen of light calisthenics immediately following his "All-American breakfast" of a cigarette and a cup of coffee. He was more reflective with a UPI reporter during that same period: "We made mistakes, a lot of them, on the previous launches," he admitted. "But nothing should go wrong up there now."

Certainly by the time Nick arrived in Sioux Falls that April, he had his game face firmly in place. Ed Yost was able to discern absolutely no change in Nick's enthusiasm or his resolve: "He was gung-ho all the way."

Before heading off to South Dakota that spring, but after reviewing the voice tapes from the Strato-Jump II flight that Yost had sent him, Nick had written Yost a letter. It provides some insight into Nick's bond with Yost, and offers a peek behind the Superman façade:

Dear Ed,

Neither Janice nor I could even begin to express our thanks and gratitude to you and Charmian. It is very easy to be surrounded by crowds of people, and yet to be very much alone. But because of you both, Janice was not alone, and no one appreciates this more than I.

At home and in traveling, I've met thousands of people, but until arriving in Sioux Falls, I never met anyone I respected completely and considered a friend. Friends are plentiful, but I don't mean that kind.

I only hope that the opportunity is presented where I can at least partially repay the consideration shown to Janice and I. If there is ever anything that I can do for you and Charmian, even if it seems impossible, please let me know. I foul up on simple things occasionally, but usually can get the rough ones done.

By the way, you're not as tough as you put on. I've listened to the tape several times—your voice changes are as evident to me as are my own, and I'm not referring to the obvious ones. Charmian can probably pick yours out, but Janice and I have only been together for four years.

Received the letter and certificate—thank you.

That is about it for me now buddy—again—thank you for everything. I'll see you in a couple of weeks.

Sincerely yours,

Nick

P.S. Get rid of this letter I'd hate to have somebody thinking I'm getting soft.

Nick's conversations with NASA had continued off and on, and he was cautiously optimistic that the agency would agree to cooperate with Strato-Jump on the next flight, supplying physiological and astronomical experiments—or at least consulting on some scientific activities. In the meantime, at Nick's request, *Life* reporter Roger Vaughan called on his connections at Princeton

University to try to interest world-renowned astronomer and Princeton faculty member Martin Schwarzschild in the research possibilities of the project. Schwarzschild, working with the Office of Naval Research in the late 1950s, had overseen some of the world's most successful unmanned balloon-borne astronomy activities, including a stratospheric flight with a twelve-inch reflecting telescope and another with a thirty-six-inch infrared telescope. But like many prominent scientists involved in high-altitude research, Schwarzschild had shown mostly disdain for missions whose first priority was lifting human cargo.

That March Nick made the familiar drive from Brick Town up to Union City to visit his parents one last time prior to leaving for Sioux Falls. While he was in the old neighborhood, he walked down to the corner ravioli shop that his old buddy Jim Lagomarsino had bought. Nick and Jim had been close, but their lives had gone different directions and they hadn't talked in several years. And Jim could sense that Nick needed a sympathetic ear that day. A three-hour conversation ensued in which Nick told the story of his recent life, much of it surprisingly unfamiliar to Jim. Nick knew that Jim was devoutly religious, and Nick revealed that he himself had recently begun attending Mass again and receiving Communion. As the evening waned, he talked about his close relationship with Janice and his daughters and mentioned that he'd purchased some life insurance to make sure they'd be taken care of in the event that anything happened to him.

"Gee, Nick," Lagomarsino remembers saying, "you sound like you're ready to die or something." It wasn't the kind of thing anybody was used to hearing from Nick Piantanida.

"Well, Jim," Nick replied, "I've got to admit that I've given that possibility a lot of thought lately. And you know, if it comes—I'm ready."

Jim remembers thinking what a strange conversation it had been. He found the experience disorienting, almost as if it hadn't really been Nick.

It was the last time the two would ever see each other.

Even as he concentrated on logistical and technical issues for Strato-Jump III, working through another long list of proposed changes that would need to be completed prior to a follow-up flight, Nick kept busy on the PR circuit. Bob Perilla helped arrange a number of radio call-in shows and TV news interviews in the New York/New Jersey area. Bamberger's Department Store installed Nick's portrait in their "Names in the News Gallery." And Nick continued his visits to schools, hospitals, and public functions that offered him the opportunity to talk about his dream, something at which he was by now not only well practiced but by all reports masterful.

Dr. Leonard Thompson, who had served as Strato-Jump's project doctor for the Sioux Falls flight in February, contacted Nick to confirm that he was eager to participate again and to request that Nick reserve a room for him at the Town House. Thompson also proposed coauthoring an article with Nick for an aerospace medical journal, suggesting that such a piece would help solidify Nick's credibility as someone who was serious about aeromedical research and that it might pay dividends if he and Istel elected to keep SPACE, Inc. going after the big jump.

On March 9 Raven Industries submitted a proposal for a balloon, instrumentation, and launch services for a third Strato-Jump flight. The operational plan remained more or less the same, as did the roster of Raven personnel who would participate. The balloon would again be a 5,000,000-cubic-foot, 0.75-mil envelope built to specifications identical to the previous one, and Ed Yost would again be in charge of operations on the Raven side. Raven's quote, $14,671, was about the same as for the February flight. But the proposal included some scheduling restrictions. Much of the Raven crew that had worked so well together on Strato-Jump II was already committed to a summertime program in Canada that would begin on June 1. If Nick could not manage to launch by the last week in May, Strato-Jump would have to be mothballed until September. The target launch date had been set for May 3, and unless they were hit with significant unforeseen delays, everyone felt confident that they could get a flight off in time.

Nick's last days of jumping at Lakewood prior to heading west to complete the mission of Project Strato-Jump occurred on the sunny weekend of April 16 and 17. Lakewood was not only Nick's home drop zone and the place where he'd made 95 percent of his jumps, it was also the birthplace of the dream. For Nick the little airport in the pines with the sea salt in the breeze provided a safe harbor in which to recharge and refocus between Strato-Jump attempts. There were no funding problems or logistical worries at

Lakewood. There were no expensive balloons or contractors, only friends and friendly skies.

He made four jumps on that Saturday, each from 7,200 feet. On Sunday his first descent was again from 7,200, but his second and third jumps were quite a bit higher, from 10,000 and 15,000 feet respectively. The third jump featured a seventy-five-second delayed opening. The final jump was from the relatively low altitude of 5,200 feet. Lakewood's director, Lee Guilfoyle, who only a few days earlier had made the first demonstration jump of a modern gliding parachute for military officials at Wright-Patterson Air Force Base, witnessed the day's activities and placed his signature in Nick's logbook. It was the last time Nick Piantanida would fall through New Jersey airspace.

Nick would jump from airplanes only twice more before his record attempt on Strato-Jump III, both relatively low-level test jumps made at Bill Jolly's request from Sheldon, Iowa, on April 24.

Nick's very arrival in Sioux Falls that spring signaled that the atmosphere surrounding the third attempt would be different. The Piantanidas landed in royal style at Joe Foss Field in a sleek private jet that was owned and piloted by Brig. Gen. Dick Lassiter, a friend of Barry Mahon's and one of Nick's newest acquaintances and financial contributors. Lassiter's name was a late addition to the official roster of Strato-Jump personnel for the third flight. Other newcomers who had gravitated to Strato-Jump by this point, individuals who remained on the periphery and appear nowhere in the project documents or media coverage, are rumored to have represented observers from both NASA and the U.S. Air Force, which spawned speculation that the government was now secretly backing the project. Nick stepped down onto the ramp in a tailored coat and tie looking like a high-powered businessman with an elegantly appointed wife on one arm. He was clearly reveling in the role of celebrity adventurer and had decided to play it for all it might be worth.

In the two weeks leading up to the launch date for Strato-Jump III, Nick held a series of standing-room-only press conferences at the Raven plant. The media turned out eagerly to hear him hold court, and he gave them their money's worth. There was really nothing to it, he would explain over and over with a mischievous grin. The flight and free fall back to Earth would require a mere three hours, and since the launch was set for dawn, he promised the reporters that he'd be back in time for 11:00 A.M. Mass.

In the evenings of the days leading up to the launch, Nick and the members of his team, as well as a rotating collection of media types and local bigshots looking for diversion, would sit drinking and smoking cigars in front of the fireplace in the Black Watch. They were all there to hear Nick expound on his exploits and to have him favor them with his attention. If they could offer publicity or had something of value to donate to the project, Nick made sure they went away happy. But for a few members of the crew, the ambience of the Strato-Jump mission had clearly changed. To some, things just felt too easy this time around, too comfortable, as if all of the difficult problems had been solved. There was a lingering sense that with the country in transition as well—some of the crew's old parachute buddies from the early days at Lakewood were now dying in Southeast Asia—that the significance of a high-altitude parachute jump seemed to have been knocked down a notch or two. They wondered if anyone was paying attention anymore. Even if the third flight was successful and Nick managed to set the world free-fall record, would anybody out there still care?

GEARING UP

There was furious discussion about the numerous changes, more than three dozen in all, that were being implemented for the third flight. Nick and Ed Yost had combed through the data and had reviewed nearly every aspect of the previous flight, and modifications had been ordered for all facets of the operation.

Due to the failure of the oxygen line's quick-disconnect, an entirely new style of fitting was being provided by Firewel, with a backup disconnect mechanism should the main system fail. Crucial instrumentation aboard the gondola had been repositioned for easier access. And, at Nick's insistence, the notorious $1.25 wrench would now be standard equipment.

The bulky glove/mitt combination that had made it so difficult to manipulate controls and couplings had been redesigned. Nick asked the David Clark Company to provide the pressure gloves one size smaller and decided to eliminate the electrically heated mitts altogether, adding low-resistance heat tape to the exposure mitts to compensate for the lost thermal protection.

The exposure mitts were also fitted with zippers and lanyards for easy removal under pressurized conditions.

The only real problem Nick had experienced with cold had been from the knees down. His torso and arms had been, for the most part, quite comfortable. The cold in the lower extremities was blamed on the furniture-pad curtains, which had been left loose at the bottoms and had flapped constantly during the ascent. Securing the curtains along the bottoms of the doors, they believed, would take care of the cold legs and feet.

As per Bill Jolly's recommendation, the ripcord on the drogue chute was lengthened and rerigged for easier manipulation. And a number of changes were made to the photographic equipment onboard the gondola, including accommodation for a telescopic camera if NASA or anyone else could be convinced to supply one.

Nick had never needed the jump tape recorder that was to have captured his comments during free fall, and it's just as well. That recorder had been overinsulated and was discovered to be partially melted when it was retrieved from the pressure suit following the flight. Nick suggested a new covering for the device constructed from the same material from which his thermal socks were made. He also wanted a smaller barograph, a small wrist-mounted still camera, and a motion-picture camera compact enough to fit in his jump rig.

Not all of the planned improvements would be ready for a spring launch. Nick had been disappointed on Strato-Jump II when he had been unable see stars in the black sky all around him. Reflected sunlight from the giant balloon envelope had polluted the skyscape and prevented any celestial observation. The Excelsior balloon had done the same thing to Joe Kittinger. At Yost's suggestion, Nick contacted an astronomy professor at Rice University who referred him to an individual at NASA's Manned Spacecraft Center in Houston who had been at work on the problem of how to observe stars during daylight hours, but nothing came of it. So Nick had worked on his own to produce drawings for a solution he called an "observation tube," which was essentially a long opaque plastic telescope—without lenses—that would run from the gondola up through the center of the balloon. But plans to implement the idea simply fell off the bottom of the wish-list as launch day approached.

Other improvements Nick had envisioned proved to be impractical, too expensive, or both. These unmade changes included an entirely new, lighter-weight gondola; a drop bar similar to those found on amusement-park rides to replace the seat belt that had proven difficult to manipulate on the previous flight; and a partially transparent aluminized net curtain to replace the furniture pads on the interior side of the doors.

In addition to the hardware improvements, Nick insisted on one addition to the preflight testing regimen: All onboard operations were to be fully rehearsed using the actual onboard equipment with a fully inflated pressure suit. This time nothing would be left to chance.

Based on long-range forecasts provided by Dr. Frisby, the planned trajectory of a May flight would result in a bailout point in the vicinity of the little town of Rock Rapids in extreme northwestern Iowa only thirty miles from Sioux Falls. While winter stratospheric winds in the Northern Hemisphere blow west to east along with the surface winds, stratospheric patterns reverse in summer and blow generally east to west. Frisby's predictions for early May anticipated the weakened tail-end of the winter pattern. This simplified things because the proximity of the projected landing site would eliminate the need for a downrange communications station, meaning Ed Yost could remain at Joe Foss Field for the entire flight and direct operations on site. This was a major advantage. And because there had been some confusion on the previous flight with too many different voices coming at Nick on the radio (at least six different individuals had transmitted instructions or questions to Nick over the course of the flight), it had been decided that only two people would be authorized to transmit: Yost and Darrel Rupp. It was also agreed that Rupp would handle the bulk of the ground-to-air voice communication. In his oral postflight report Nick had mentioned that these communications restrictions were to be strictly observed. "This includes Istel," he had made clear. "There was a fairly constant line of jabber coming out of ground control at Estherville."

Nick also decided that, in contrast to the free-form conversations of Strato-Jump II, a more rigid communication format would be adopted. At each 10,000-foot interval during the ascent, he would provide the ground crew a complete readout of his instrument panel in order to avoid the constant requests for readings.

The weather would apparently not be an issue this time around. It was a lovely spring in South Dakota: daytime temperatures in the mid-seventies, nights in the forties and fifties. And even when a cold front moved through on April 28, dropping the overnight temperatures below freezing and kicking up winds of 25 miles per hour, there were few concerns about launch delays for Sunday, May 1. Shortly after arriving in Sioux Falls this time, Nick requested that the launch date be moved up. "Why shouldn't I go on the Sabbath?" he asked. "After all, that's God's day. And I might as well do it on May Day. That's the Russians' big day." The traditional International Workers' Holiday in Moscow that year would take as its theme U.S. "barbarism" in Southeast Asia. The Soviet defense minister declared publicly that American troops in South Vietnam "were using the most cruel and barbaric means of annihilating people." A float towed through Red Square depicted a bandaged Vietnamese child beneath the slogan: "Bring the murderer to account!"

It is not clear whether anybody in that expectant Sioux Falls crowd gave even a fleeting thought to another significance of "May Day": the international distress signal for sailors and aviators.

Over late-night drinks in the Black Watch, Dick Kelly presented Nick with a green rabbit's foot that he'd carried during his own navigator training a few years earlier. But even though Kelly had seen Nick originally as a man hell-bent to kill himself, the rabbit's foot was offered in laughter and gratitude for interesting times rather than as a hedge against disaster. Nick's own confidence, meanwhile, seemed to most observers to grow as the prelaunch days dwindled, his enthusiasm and vitality spilling out and drenching everyone who came into contact with him.

As before, most of the preflight action took place in the flight operations area of the Raven building downtown and at the National Guard hangar at Joe Foss Field. This time three chase planes would be employed: Raven's Cessna 180 would again be piloted by Frank Heidelbauer who had agreed to take along backup flight surgeon Dr. Claude Mattingly and Dick Wagaman, who was the ranking official observer from the Parachute Club of America; Raven pilot Gerald "Dusty" Rhoads would fly the company's Cessna 185 and would carry Bill Jolly, Barry Mahon, and Dan Quinn; and another Raven pilot, Al Tomnitz, in a borrowed Piper Super Cub would be accompanied by Dr. Thompson.

The Raven group once again assembled and prepared the massive balloon while the Strato-Jump crew put finishing touches on the gondola. There was a new look to the craft's exterior, and a curious but eye-catching new moniker. A fresh red and white placard glued to the new styrofoam panels read:

S.P.A.C.E. inc.
STRATO JUMP III "JADODIDE"
N30128
EXPERIMENTAL

"JADODIDE" was Nick's amalgam of the first two letters of the names of his wife and daughters (Janice, Donna, Diane, Debbie).

On the afternoon of April 29 the refurbished gondola was loaded on a truck at the Raven plant. Early the next morning, Saturday, the truck was moved the few miles to the National Guard hangar at Joe Foss Field and unloaded. Marvin McCall, who had become an instantly recognizable figure at the Raven plant because of the fact that he always reported for work dressed entirely in blue, packed the Dave Clark Special and the other garments Nick would wear into a special container and packed a separate case with items that would be tucked into various pockets on the suit prior to launch: a fresh stack of cover letters wrapped in plastic and including the earlier letters to Janice and Donna (McCall had been given the assignment of having the covers canceled at the Sioux Falls post office), a roll of nylon tape, flotation gear (in case Nick landed in a lake or a river), extra boot laces, an assortment of rubber bands, cigarettes and matches, a list of important phone numbers, and five dimes. One additional item, a good-luck charm, would be slipped into a suit pocket just before launch: Janice's wedding ring. It had accompanied Nick on both the previous flights.

At 2:00 P.M. Nick held his final press conference; at 4:00 final briefings for all personnel were conducted, first by SPACE, Inc. and then by Raven. Even mundane details such as seating arrangements in the recovery aircraft were reviewed. A discussion on emergency procedures covered contingency plans in the event of serious injury or fatality. At 8:00 Dr. Frisby presided over a final weather briefing that left the Strato-Jump crew in high spirits. She was forecasting clear skies and light winds, and all systems were go. With the exception of an ailing Jacques Istel, who seemed to be complaining about everything by this point, it would have been difficult to find anyone in town willing to bet against the success of the next day's flight. Nick would rise up to peak altitude (they had already proven this could be done), and this time he would disconnect his onboard oxygen and leap from the gondola into the record books. Nick's skill as a skydiver would take care of the rest.

As one member of the crew put it: "How hard could it be? You go up and jump out and set the record."

Nick and Janice had gone to church early that afternoon and Nick had made confession. That evening, the core group gathered in Barry and Clelle Mahon's Town House suite: Nick, Janice, the Mahons, Ed and Charmian Yost, Dan Quinn, Phil Chiocchio, with a handful of others coming and going. But unlike previous preflight nights, there was a distinct—and, to some, a troubling—lack of urgency in the room this time, as if the next day's mission was already a *fait accompli*. In spite of Nick's repeated efforts to keep the focus on the project, the discussion kept veering away to all the myriad other things on people's minds: the war in Vietnam, politics, fashion. Phil Chiocchio recalls Nick looking uncomfortable and making eye contact with him as the conversation turned to—of all things—women's dress shoes. Moments later, Nick and Phil were outside and walking the darkening streets of downtown Sioux Falls. They ended up at the counter of one of the town's best diners, the Hamburger Inn on Tenth Street.

"It was a weird evening," Phil recalls. He had never seen Nick so quiet, so preoccupied. As if, alone among his team, he harbored some deep concern about the events that would unfold once the sun rose, a concern he had managed to keep hidden from almost everyone but his wife.

Many of the Strato-Jump guys celebrated hard that night, as had been their custom while in Sioux Falls. A group of five or six went to Jim Lee's for ribs. Nick returned to the motel and had a couple of quiet conversations in the Black Watch before he and Janice turned in for a few hours of restless sleep.

"Emergen[cy]"

The Raven launch crew was the first to arrive at Joe Foss Field on May Day morning. They'd reported to the company plant downtown at 2:15 A.M. and had left for the airport a little after 3:00. Nick's wakeup call had come at 2:30 and he and Janice had left the motel parking lot at 3:15. Strato-Jump personnel were scheduled to arrive no later than 3:30, but not all of them made it on time. The layout and inspection of the pressure suit and all associated gear was scheduled for 3:45 with the suit-up and chute-up procedures to follow. But by the time Ed Yost got to the field shortly before 4:00, he found Nick, already in the pressure suit but with his helmet visor open, sitting on the tailgate of a stationwagon talking casually with Janice and Jacques Istel. According to the prelaunch checklist he should still have been in the suit-up area in the National Guard administration building as the inflation of the balloon began.

"What the hell's going on?" Yost asked, his head cocked to one side and his big eyes seething. He didn't like anything about the way this day was starting.

Nick explained that part of his crew had not yet shown up on the flight line. Presumably they had overslept. They'd been called and would arrive shortly, he promised. Yost was openly appalled, but Nick's spirits appeared once again to be sky high as he did what he could to lighten the mood.

"No problem, Ed," he said, attempting to defuse any worries. "We're okay. We're okay." Still, the situation was urgent. Once inflation of a big plastic balloon begins, there is no turning back; a polyethylene balloon can be inflated and used only once. And once the gas bubble begins to form, the fragile envelope cannot be held on the ground indefinitely. One can only guess at the pressure Nick was under to make sure that nothing interfered with the launch that morning, and what it might have taken to have convinced him to waste the expensive balloon and postpone the mission for another day.

Janice listened impassively. Her trust in Nick's solo act was supreme. She believed in it. If Nick said there would be no problems, the case was closed as far as she was concerned. Besides, neither she nor Istel—nor even, on that morning, Ed Yost—had the power to talk him out of this flight. The weather was perfect, the balloon would be ready soon, the gondola with its new and improved oxygen disconnect was rigged, and if Nick Piantanida had any doubts or felt at all cornered by his own sense of destiny, he did a masterful job of hiding it.

At 4:30 the familiar whine indicated that the ritual of the balloon inflation had begun. At approximately 5:20, having returned to the administration building and completed the chute-up, Nick reappeared, waved to the assembled crowd, and took his seat in the gondola to begin his onboard prebreathing of oxygen. About twenty minutes later, he gave the thumbs-up signal to Yost.

Nick's voice came across the monitors in the communications van. "You ready to release, Ed?"

It was Darrel Rupp who answered. "Nick, this is Darrel in the van, ready for a final radio check." Rupp's would be the single voice of Sioux Falls ground control that morning.

"Roger," Nick replied. "One-two-three, one-two-three."

"Okay," Rupp said. "You sound good."

Some four hundred locals, a much larger contingent than had turned out for the previous attempt, had braved the subfreezing temperatures and driven out to Joe Foss Field at dawn.

Janice had stuck unusually close to Nick during the preflight sequence, but now she backed away from the gondola and joined the onlookers smoking and drinking coffee. She was a veteran now, too, and she knew the routine. "He doesn't need a woman to get in his way," she whispered to Jacques Istel. Charmian Yost hugged Janice and held onto her for a moment as technicians and cameramen swarmed around the gondola.

"Oh God, he has to make it this time," Janice said to no one in particular. Other than Nick, she was the only one there who knew precisely just what had been invested in this attempt—not only money, but the years of research and preparation, the sacrifices they'd made, the strain on the family. It was the moment of truth for them both, and Janice could feel the pressure as surely as if she'd been suited up and seated in the gondola.

"Okay," Nick announced. "The timer's set. Everything is all armed. See you in about three hours."

At 5:45 A.M., just as the sun's color broke on the horizon and roughly fifteen minutes ahead of schedule, the elongated teardrop of clear plastic plucked the gondola from the launch vehicle and yanked it cleanly up into the five-mile-per-hour wind. Strato-Jump was aloft for the third time. And this time the launch was so smooth that Nick was unable to distinguish the instant of liftoff.

"Tell Keuser I owe *him* a steak tonight," Nick said.

As the gondola lifted away, the crowd applauded, gloves and mittens muffling the effect. Nick triggered Mahon's film camera and captured a brief pan shot, sweeping across the hangars and runways of the airfield. The balloon blew north and west of the airport for the first 2,000 feet of the ascent and then drifted due east and back across the city of Sioux Falls precisely as predicted by Dr. Frisby.

"Make the rate of rise about 950 [feet per minute]," Nick reported.

"Nick," Rupp suggested, "rather than waiting until 10,000 feet, do you want to check your reset timer now?" At each 10,000-foot interval, Nick was to perform a brief series of checks and give a readout of the instrument panels. At the same time, he would reset his fifteen-minute timer and verify that all automatic mechanisms were properly armed.

"Roger," Nick said. "Reset timer: reset."

"Roger, and check your arm switch to make sure you didn't shut it off."

"Roger," Nick said. "Arm switch on, arm camera on." Nick proceeded to read off his instrument checks. "Make the time to be five fifty-two. Pulling 23 volts. Six amps. Outside temperature looks to be about 8 below. Inside is zero."

"Roger, Nick," Rupp said. "Your altitude is reading about what?"

"Now I'm at . . . coming up on 11,000 feet. Two-hundred fifty [feet] shy."

"Roger. Did you check to see that your arm switch is still armed?"

"Roger. Arm camera and the cut-off timer both still on."

"And your reserve automatic opener is activated?"

"Yes sir," Nick answered.

"Okay, thank you."

As the balloon rose, time and altitude marks were announced on the ground and posted on a big table positioned outside the communications van. Ed Yost monitored the rate of ascent, shooting for 1,000 to 1,100 feet per minute. The strategy, once again, was to get Nick quickly through the tropopause and into the stratosphere where the chances of a balloon failure were greatly diminished. Nick's reported rise rate had been a little slow, so now that the instrument check was complete, Yost asked for a ballast drop.

Rupp relayed the request to the gondola. "On your next minute coming up, Nick, would you drop out two minutes of ballast? Two minutes on your next minute." It was necessary for the operations team to know precisely when each ballasting sequence began and ended so that accurate rise-rate calculations could be made.

"Roger," Nick acknowledged. Once the two minutes of ballasting were complete, Rupp reported the result back to the gondola.

"Nick, that ballast picked you up to 1,000 feet a minute."

"Roger." Thus far, the flight's progress was close to perfection.

As *JADODIDE* ascended through the troposphere, Nick began to feel the cold.

"Closing gondola curtain," he reported. "Getting just a little bit of a temperature drop." Moments later Rupp's voice was back on line for another instrument check.

"Nick, this is Sioux Falls. We can run down our 20,000-foot check."

"Roger," Nick said. "I'm about 100 foot shy of it [20,000 feet]. Time: oh-six oh-two. Altitude: now 20,000 on the button. Outside temperature is about 18 below zero. Inside temperature: 5 below. Voltage is reading 24. Current: right on 6. Reset timer: on to zero. After reset check, camera switch on. Cut-down timer on." His voice was calm and businesslike, sounding almost bored at times as he relayed his progress to the ground. These conversations always reminded Janice of the chatter heard on televised NASA missions.

"Roger. Thank you."

Nick's breathing was relaxed and regular. His respiration rate had held steady since the launch at about seven and a half breaths per minute. Every minute or so, Rupp would relay a question or a suggestion from Ed Yost. This dialogue was intended partly to gather specific information and partly to maintain contact with Nick in order to monitor his state of mind.

"Nick, do you have any sock or visor heat on now?"

"I'm using visor heat," Nick answered.

"Roger, thank you."

"The visor is just a little bit moist," Nick said. "I'm using the heat to dry 'er off." There was no concern in Nick's voice, but his breathing picked up just slightly and became a bit less regular.

"Make my rate of rise out to be about 1,100 feet per minute." Nick's delivery seemed to slow slightly, but not enough to cause any worries on the ground.

"Roger," Rupp said.

From time to time the Raven pilots would report their positions on the same radio frequency. Occasionally the pilots would request specific information such as Nick's last reported altitude. Nick could monitor these transmissions but rarely responded or commented.

"Ed," Nick announced, ready for another instrument check, "we're coming up on 30,000 feet."

Again, it was Rupp that responded. "Roger. Go ahead, Nick."

"Time is oh-six thirteen. Outside temperature is 31 below. Inside is 4 below. Voltage is reading 23 and a half. Amps are pulling 6 on the button. Resetting timer for 15 minutes."

"All right, Nick. At this time we'd like to have an oxygen check."

"Roger," Nick said. "Oxygen reserve [tank] reading about 1,800 pounds [per square inch]. The main [tank] is on 1,700."

"Thank you. And your reset timer is still armed?"

"Reset timer is armed," Nick said. "Camera is armed."

"Roger, thank you."

A few minutes later and the gondola had passed through 35,000 feet, the altitude at which the Dave Clark Special was designed to inflate.

"Nick," Rupp asked, "have you started to pressurize yet?"

"Roger," Nick said. "We're holding steady at about 34,000 [feet]"—indicating that the suit was maintaining the pressure equivalent of 34,000 feet above sea level as the balloon continued to ascend. Now it was time for yet another instrument check.

"I'm at 40,000 feet." Nick's slightly bored inflections indicated that these readouts were not his favorite task, which may explain the apparent carelessness of his next transmission. "Time is oh-six seventeen and a half." Even though Rupp did not question or dispute this reading, it was undoubtedly in error. At 30,000 feet Nick had reported the time as 6:13. To cover another 10,000 feet in four and a half minutes would have required a sudden, dramatic acceleration in the balloon's rate of rise to a dangerous—if not impossible—2,200 feet per minute. Nick almost certainly misread the clock by five minutes and in reality had reached the 40,000-foot level at 6:22 and a half.

"The outside temperature is 40 below," he continued. "Inside is about 4 below. Voltage reading is holding on 23. The amps are pulling at 7. I'm resetting the timer. Cut-off timer is still on, the camera is still on."

As the temperature plunged, Nick found it necessary to keep the pressure visor's heat turned on. "I'm using a little bit of visor heat to keep this visor dry. Every time I shut it off, she starts to steam up a bit."

"Roger," Rupp acknowledged. "And how's your oxygen doing?"

"Oxygen is at 1,600 on the main and reserve is still the same: holding at about 18."

"And your suit performance, Nick?"

"Suit performance has been holding steady at 36,000 feet."

The balloon had passed well through the trickiest part of the ascent, the tropopause, without incident or comment from Nick. Now that he had made it into the stratosphere, Yost was curious about the performance of Raven's balloon.

"Nick," Rupp said, "Ed would like to know if you've noted any unusual movement or rotation of the balloon going through the trop."

"Yeah," Nick replied. "I was noticing the rotation, but the balloon didn't seem to bounce at all. Last time I felt just a little bit of a jar. This time it was pretty smooth. Uh, do I get a buzzing sound when the thermostats go on up on the cameras? I've been noticing just a little bit of a click, and at the same time I see the amps drop."

There was a pause as Rupp, Yost, and others discussed the situation.

Rupp's voice came back on line: "It's very possible, Nick. There'll be probably a little bit of an arcing as those thermostats open and close."

"Okay." Nick's voice was relaxed and confident, but his breathing had begun to pick up pace again. In the earlier stages of the flight, his rate of respiration was about seven breaths per minute. The rate was nearly double that now.

Dr. Frisby's forecast had held up generally well, but Nick's eastward progress along the Minnesota-Iowa state line was now quite a bit faster than anyone had anticipated. This was not a problem, except that the General Electric van would need to adjust its location in order to stay within range and to be positioned properly to monitor Nick's jump.

"Nick," Rupp said, "at 6:24 you were over Magnolia, Minnesota. In case the G.E. people do not hear us—which they should, through you—would you relay to them that they should begin moving east?"

"Roger," Nick said. "G.E. van: Sioux Falls radio advises you to move east. How far, Darrel?"

"Nick, they should go probably at least through Worthington." Worthington, Minnesota, is some sixty-five miles due east of Joe Foss Field and about ten miles north of the Iowa border.

"G.E. van: Sioux Falls advises that you move at least to Worthington."

Nick's rate of rise had slowed again. So, as he passed through 46,000 feet, Yost ordered another two-minute ballast drop, which brought the rate back up to 1,000 feet per minute. A couple of minutes later and they were ready for another instrument check.

"Comin' up on 50,000 feet," Nick said. His voice was more subdued by this point, but still strong. His breathing was regular, if noticeably faster and a bit more labored than it had been earlier. "Time: oh-six thirty-four. Outside temperature is 45 below. Inside: 4 below. Voltage reading is at 23. Pulling 6 amps. Resetting timer. Timer reset. Camera arm is on. Cut-off timer's on. Still using visor heat."

"Roger, Nick."

"Suit pressure is now back to 35,000 feet. Oxygen is good. Reserve is still holding same. The main is at . . . looks to me to be about 1,300. Yeah, it's 1,300 psi."

"Roger," Rupp replied. "Sounds like your inside temperature is holding real stable."

"Oh yeah," Nick said. "It's beautiful!" This was good news since the balloon was now in the coldest region of the stratosphere. The outside temperature could be expected to stabilize and gradually rise during the rest of the ascent.

"Do you still have your curtains closed?"

"Roger," Nick confirmed, all the boredom gone from his voice. "I've been taking looks out. There's an overcast and I can barely make out a little bit of the ground below it. I'm now facing east, right into the sun. Can't see the ground. Probably coming around to the west. Oh, now I can . . . still facing east, but I can make out the ground slightly."

"How much rotation are you getting now?"

"Right now, not much at all. Up until about a minute ago I was getting a slow rotation—not too much. It seemed to be going . . . rotate about 180 and then back. Now I'm rotating toward the north. Now the ground is . . . oh, it's quite visible. It's just very light in a few spots, all the way up to the . . . now I'm rotating to the east." Nick seemed to enjoy the opportunity to describe something other than his instrument panels. "Comin' right back into the sun," he added.

Rupp reported the balloon's position: "You're eight miles east of Adrian [Minnesota] right now, Nick."

"Roger." Nick had studied the maps and knew precisely where he was. "Kinda got there a bit ahead of time."

"Yost says your wind should be decreasing pretty quick now."

"Roger."

A minute later, Rupp confirmed Yost's prediction. "Nick, Dick Keuser said that you've slowed down and are moving very little now." Nick was still ascending at about 1,000 feet per minute, but his progress across the ground had stalled. Anxious to put some distance between Nick and the necklace of small lakes surrounding Worthington, Yost decided to bump the rate of rise up just a bit in the hope of finding some more wind.

Rupp relayed the ballast request: "Okay, Nick. On your next minute, would you drop another one minute of ballast?"

"Roger."

At this point, with the balloon at an altitude of nearly 57,000 feet—in another five or six minutes it would hit the Armstrong Line—Al Tomnitz, circling far below in the Super Cub, came on line to offer a visual report of Nick's position.

"Ed," Tomnitz said, "I put him right over Worthington. About three miles west of Worthington at this time."

"Roger," Darrel Rupp acknowledged. "Thank you, Al."

" . . . goin' on the ground there," Tomnitz added, apparently activating his microphone in midsentence. He may have been commenting on the General Electric van that was rolling into the Worthington area at about that time.

Then, several moments after Tomnitz's comment, the team on the ground in Sioux Falls heard a sudden and unfamiliar sound. Darrel Rupp, Ed Yost, Janice Piantanida, and others standing within earshot of the monitors that broadcast the transmissions from the tiny microphone in Nick's helmet heard it. To most of them, it sounded like a short blast of wind. Yost said he heard a "whoosh." Others described a "hiss." The brief whoosh was immediately followed by Nick's voice. But nobody could make out what he'd said, if in fact he'd actually said anything.

"What was that?" Yost asked quickly.

"Maybe he sneezed," Janice said, turning to face Yost.

Several seconds after they heard Nick's voice, Darrel Rupp asked for a clarification.

"What was that, Nick?"

The reply came almost before Rupp could get the question out.

"E*mer*gen . . . !"

Nick had started to say "emergency" but had either been cut off or had for some reason been unable to complete the word. Whatever the case, the radio transmission from the gondola had gone dead at that instant.

"Emergency! Cut him off!" Yost barked. It was a gutsy call. With no time to investigate, and no real assurance that they were dealing with anything more than communication problems, Yost ordered the flight of Strato-Jump III—and the plans and hopes and money it represented—terminated. There would be no free-fall record attempt this May Day.

Instantly, Darrel Rupp hit two pickle buttons on the console that triggered a radio signal that blew the huge balloon free of the gondola. Nick—in what condition, nobody knew—would now fall once again beneath the cargo parachute. It would take twenty-six agonizing minutes for *JADODIDE* to reach the ground.

Janice, fearing the worst, all color draining from her face, spun away from the booth, her hands trembling.

Yost cradled her with one arm and called to his wife who was standing just outside the van. "I want you to go home and get a change of clothes and an overnight bag," he calmly instructed Charmian. "You're going to meet Janice at the hospital."

Frank Heidelbauer, Al Tomnitz, and Dusty Rhoads were already in the air and converging on Worthington. Only minutes later, Dick Kelly was in his car and speeding off for Interstate 90 with Janice beside him, scarf cinched even more tightly than usual beneath her chin. A badly shaken Jacques Istel rode next to Janice, with Istel's nurse and a member of Mahon's film crew in the back. Janice was in shock but was still cognizant enough to be aware of Istel mopping the sweat that seemed to be pouring in sheets from his face and neck. During a brief stop at a traffic light on the outskirts of Sioux Falls, Istel's nurse handed him a syringe and Janice watched him inject himself in the forearm.

"I think the stress of it just really did something to him," she says. "He must have been very, very ill."

Once on the interstate, Kelly fell in behind a hastily arranged Minnesota Highway Patrol escort going full blast with sirens wailing and lights flashing.

The guessing games began before the gondola had even hit the ground. There was speculation that the air may have rushed out of Nick's lungs, keeping him from vocalizing the final syllable of "emergency."

Darrel Rupp, an experienced communications technician, had another theory. "His communication system and his power during the flight," Rupp explained, "was provided from the gondola through a major fifty-pin electrical connector, an umbilical cord that plugged into the gondola. And because his communication shut off instantaneously in the middle of the word, it was like he knew something went wrong and he was quick reacting to pull that umbilical and he was ready to go out. So he had started to say 'emergency,' you see. He was headed down."

Worthington

The gondola came to ground in a barren field some twenty miles east of Worthington, Minnesota, about four miles southwest of the town of Lakefield. The balloon envelope landed just a few miles away. Raven pilot Al Tomnitz got the Super Cub on the ground at almost the exact moment the gondola touched down and taxied across the field to within a few yards of the landing spot. Dr. Thompson scrambled from the plane with his medical case in hand and ran to the gondola which was tipped onto its side. Nick's personal parachute had deployed at about 7,000 feet during the descent as designed, and when Thompson dragged Nick out, the piles of nylon spilled out alongside them. By the time Tomnitz had climbed down from the pilot seat, Dr. Thompson had already removed Nick's helmet and was in the dirt with Nick's head in his lap. Heidelbauer had since landed the Cessna 180 nearby and had also arrived at the gondola.

It was obvious to all of them that Nick had been badly injured. At first, he was writhing and moaning on the ground, and Thompson, concerned that he would choke, had pinched Nick's tongue to his cheek and barked at Tomnitz, "Hold this!" Seconds later, Thompson pierced the tongue and cheek with a needle and fixed them with a clamp before performing mouth-to-mouth resuscitation.

"That sure shook me up," Heidelbauer said. Tomnitz, a tough, unflappable pilot, concurred: "It shocked the hell out of me."

Kent Morstad was living in Worthington where he worked for a funeral parlor that also ran an ambulance service. He was relaxing at home that morning when he got a call about a balloonist that had crash-landed nearby. An

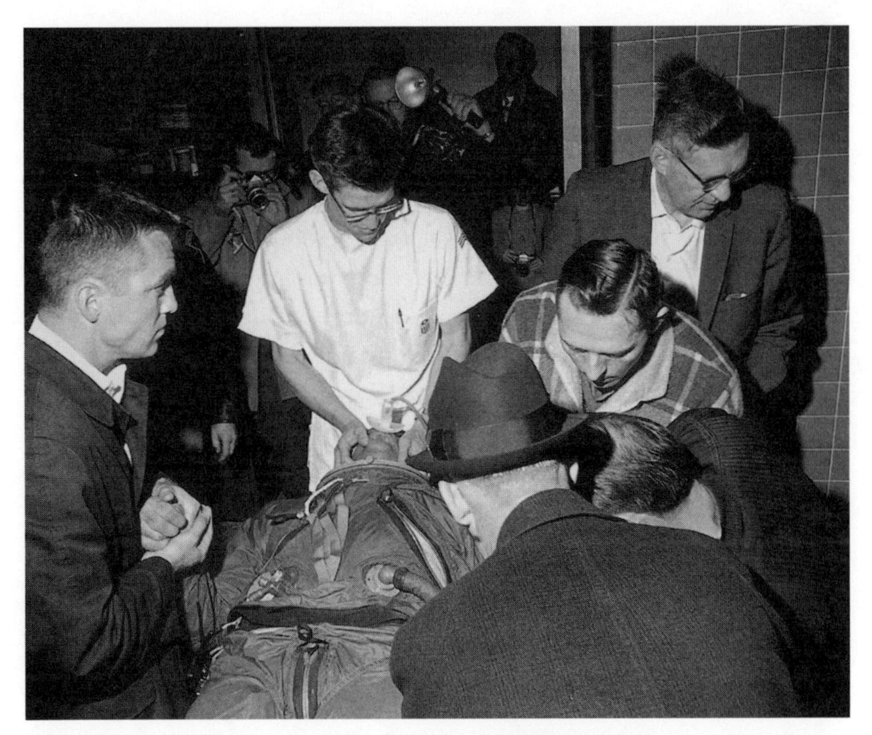

The scene as Nick, unconscious, arrives at Worthington Municipal Hospital; Kent Morstad holds Nick's hand at left. (Courtesy TimePix)

ambulance driver picked Morstad up and they raced to the designated location. When they reached the edge of the farm where more than a dozen vehicles already lined the road, they could see people and a pair of airplanes in the field, but it took a couple of minutes for them to find a break in the surrounding fences that would allow them access to the site. The balloon had come down about fifty yards from the road.

One of the cars on the shoulder of the country road belonged to Lucy Stefan, who had driven from Minneapolis down to Sioux Falls for the launch with her children. The Stefans had followed both of the Raven flights, and Lucy had wanted the kids to be able to see Nick on his day of triumph. Now they were parked at the landing site and were witness instead to horror.

By the time the men got Nick onto a stretcher and into the ambulance—an operation complicated by Nick's size and the bulky pressure suit; six men bore the stretcher to the ambulance—Kent Morstad observed no body movement and heard no sounds from the unconscious man. Dr. Thompson jumped in the ambulance. The high-speed trip to the tiny Worthington Municipal Hospital took about fifteen minutes.

Neither Morstad nor Thompson was prepared for what they would encounter at the hospital. The scene was frantic—total chaos. Strato-Jump and Raven crew members were arriving every few minutes from Sioux Falls, and the press, both newspaper and television, was already swarming. For a few minutes Janice wandered the main hallway bewildered, reporters following her with cameras, badgering her with impossible questions. Morstad could think only of vultures picking over a carcass. When members of the local sheriff's department began to arrive, they did their best to clear the crowd, but for a while at least it was a losing battle.

The hospital staff refused to allow Janice into the emergency room, but a chair was brought in so that she could wait in relative privacy just outside. Dick Kelly gave her a cup of coffee and fortified it with a dollop of whisky from a flask. "Janice," he said, "I want you to drink this. The doctor says it's okay."

She tried to drink, but her hands were shaking too badly to hold the cup. She could hear Nick moaning in the next room. *Why is he moaning like that?* she wondered. *Is he really in pain, or is he just trying to breathe? Why won't they let me see him?*

The first big problem was getting Nick—still unconscious—out of the pressure suit. An airway tube had been inserted down Nick's throat and doctors were unwilling to extract it in order to facilitate removal of the suit. They ended up having to slice the suit fabric just below the metal helmet ring in order to strip it off. Unfortunately, there was no local expertise in Worthington to deal with an accident like this. Dr. Thompson had performed a tracheotomy to ease Nick's respiration and had prescribed medication to treat pulmonary edema, but beyond that there was some confusion about the proper first aid to administer.

Meanwhile, just outside the hall that had by now been mostly secured by local law enforcement, the shouts and shoving continued unabated. No one knew the extent of Nick's injuries or had any idea what the prognosis might be. The cause of the accident was a mystery. If Nick had for some reason lost pressure in the stratosphere, it was hard to know what to expect. High-altitude decompression was an extremely rare phenomenon, and no one on site had much real experience in the matter. It was an especially frustrating time for the crew members and contractors. As a pained Ed Yost observed, "You just couldn't get any information."

The glorious morning had soured into nightmare. Reporters prowled the perimeter of the little hospital. Next to Janice, who had finally been able to join Nick, and the doctors, Ed Yost was the one the writers most wanted to talk to. And yet Yost knew little more than anyone else. The survival system must have malfunctioned, he guessed. Perhaps something on the helmet failed? Or perhaps the suit had been punctured or an oxygen line had been dislodged? Who knew? As frustrating as it was for all concerned, there simply were no answers available. Nobody could get a word from the medical staff.

Frank Heidelbauer recalls his short solo hop back to Sioux Falls in the Cessna. As he was cleared to land at Joe Foss Field, the traffic controller informed him that a local television news crew from station KELO was waiting near the tower and wanted to interview him. But the normally jocular Heidelbauer was numb. He listened to the reporter's urgent questions but found nothing to say and could answer only with silence.

Bill Jolly arrived in Worthington not long after the ambulance, and he and Dick Kelly made it their first order of business to insulate Janice from the press. Jolly had phoned his superiors at Pioneer Parachutes to inform them of the situation, and they had told him: Whatever you do, don't leave Janice.

"Mr. Kelly and Bill Jolly," she says, "they saved me. Without them . . . I don't know." The heroes of the day—in addition to Dr. Thompson and the Worthington medical staff who were doing everything they could—were Jolly, Kelly, and the Yosts, who appreciated that Janice, at age twenty-three, was suddenly very much isolated and in a cruel free fall of her own. They stepped in to provide what safety net they could for this woman who had so totally defined herself in relation to her dynamic husband and who had relied on him for practically everything.

At one point someone yelled down the hall that Janice had a telephone call, that it was her mother. She went down to the phone booth, but when she picked up the receiver, the voice of a reporter came on the line, asking, "Can you describe for our readers what you're feeling right now?"

Jolly arranged with the Worthington staff for Janice to have the exclusive use of the hospital room adjacent to the emergency room where Nick was being monitored, and that was where she spent that horrible night. Charmian Yost arrived and took the second bed to keep Janice company. But it was impossible for either of them to get any sleep. Off and on throughout the night, Nick—just beyond the wall—would begin to moan.

Early reports from the attending physician at Worthington Municipal, Dr. L. B. Dawson, were cautiously optimistic. The working assumption was that catastrophic decompression and subsequent loss of oxygen were involved. Yet in spite of the fact that Nick may have suffered the highest decompression on record—six minutes later and he would have hit the Armstrong Line where his blood would have boiled in his veins—he was still alive and his pulse and blood pressure had been stabilized. That was something just short of a miracle right there. With his great strength and iron will, it was hard for anybody to imagine that he wasn't going to snap out of it.

Nick was still . . . Nick.

6 VIGIL

It was a mild spring morning in Union City. Mass had just ended as Vern Piantanida filed out of church and felt a gentle tug on his arm. A family friend, having caught a bulletin on the radio, leaned in close: "I'm so sorry to hear about your brother." It was the first Vern had heard of the accident in Minnesota. Word about Nick had always spread fast.

Back in Sioux Falls the next morning, the front-page banner headline screamed: NICK STRUCK SUDDENLY BY CRIPPLING DISASTER. Just "Nick"—no last name needed in that community. The *Des Moines Register* went with a five-column, above-the-fold aerial photo of Dr. Thompson rushing from the Super Cub to the gondola. An inch-high banner read: 'EMERGENCY' AT 57,000 FEET! BALLOONIST BLACKS OUT, NO PRESSURE.

After watching as Nick was loaded into the ambulance and making his own perfunctory stop at the hospital, *Life* reporter Roger Vaughan returned to the Town House Motel in something akin to a trance. Janice remembers passing his room, number 218, late that day on her own devastating mission of collecting her and Nick's belongings from their room, the room they'd shared together the night before. The curtains were open and she could see

Vaughan's small, whiplike frame hunched at a desk, typing furiously. Vaughan said that his *Life* editors had given him another assignment; they wanted him to finish up the Strato-Jump piece that afternoon and catch an evening flight to London. But Vaughan told Janice that as far as he was concerned, *Life* magazine could go to hell; he wasn't leaving Sioux Falls until he could finish the story the way he wanted to write it. He worked into the night, rewriting, agonizing over the words, trying to find the right way to tell it. His account of Project Strato-Jump would be not only his tribute to a man he had grown to admire, it would be a way of coming to his own terms with the meaning of what was shaping up to be a terrible tragedy. Somewhere in that lava flow of fact and sentiment, he would eventually find the story he filed the next morning.

"One in a Million"

In the early morning darkness of May 2 Nick Piantanida was transferred from Worthington Municipal to Hennepin County General Hospital in Minneapolis where, at about 7:00 A.M., he was finally placed in a hyperbaric chamber. Some twenty-four hours had already elapsed since the accident. The physician who took charge of Nick's case, the hospital's chief of surgery, Dr. Claude Hitchcock, had been partnering with the U.S. Navy on medical treatment for divers suffering from severe decompression sickness. The hospital's state-of-the art chamber, a sixty-five-foot-long, blue-steel, cylindrical vault was designed to provide the body and brain with a pressurized, oxygen-rich environment in which to attempt to heal. In addition to the treatment of divers, hyperbaric oxygen therapy had also been proven effective for some victims of other ailments such as carbon-monoxide poisoning, radiation tissue damage, and gas gangrene. However, Hennepin General's chamber had never before been used to treat a case of high-altitude decompression. In fact, Hitchcock had thought it prudent to consult a Navy doctor in San Francisco, the head of the USN's "Man Under the Sea" project, by telephone about the proper approach to Nick's injury while the plane was en route from Worthington.

The sixty-two-ton chamber, a highly specialized, half-million-dollar installation and a far cry from the earliest "recompression chambers" that had come into existence around the time of the Civil War, was flooded with pure oxygen and pressurized to the equivalent of 66 feet below sea level, and the pressure was gradually decreased to a sea-level equivalent over a six-hour period. This was more or less standard procedure for acute DCS. Technicians monitored Nick's condition by means of a closed-circuit television hook-up and two-way communication while doctors in pressurized suits took shifts inside the chamber to attend to Nick. The procedure had been fairly effective with Navy divers, but its potential value in Nick's case was open for debate. The patient was obviously in bad shape, and Hitchcock tried to limit expectations by announcing that some degree of brain damage was probably a foregone conclusion, but when pressed he admitted that the long-term prognosis was entirely unclear. It had taken the gondola roughly ten minutes to descend beneath its parachute to an altitude of somewhere between 15,000 and 10,000 feet where Nick would have had sufficient oxygen to breathe normally. Brain damage is generally thought to occur after three minutes without oxygen.

"We're giving him every chance," Hitchcock told the press, reluctant to snuff out all hope. Apparently Nick's breathing had steadied and he had exhibited some minor movement following the six hours in the chamber.

Janice returned to Worthington from the Town House Motel, and both she and Bill Jolly were hustled out of the intensive care unit of Worthington Municipal and escorted directly to the local airfield and onto the ambulance plane. The privacy and respite from the howl of the press were welcome, but as they touched down in Minneapolis just after dawn they found a whole new contingent of newspapermen there to meet them. Janice was relieved at least that the new environment at Hennepin General seemed to have quieted Nick's moaning. That had been the hardest thing for her to endure.

"It was something you couldn't listen to for very long," she says. "Nobody could."

Hennepin General was a jarring contrast from the low-key facility in Worthington. As they had wheeled Nick into the busy hospital on a gurney, a newspaper tent over his face to provide some shelter from the prying eyes and cameras, Dr. Thompson walking beside him holding a pair of IV dispensers from which tubes ran into Nick's arms, Janice saw things she had never imagined. This wing of the hospital was a veritable showcase of exotic malady: the man in the room on one side of Nick's had damaged intestines that wouldn't heal and he had violent screaming fits every few hours. The girl in the room on the other side had too much calcium in her blood and was losing her limbs one at a time. Luckily, a young nurse, a New Jersey native, who had heard of Nick, got herself assigned to his case and became a source of comfort and optimism for Janice. The nurse said that she had worked with lots of trauma patients and urged Janice to talk to Nick, tell him he would be all right, encourage him.

So Janice talked to Nick every day, stubbornly refusing to be discouraged by the medical reports or the obvious concern of the Hennepin staff. "One time I was holding his hands," she recalls, "and I knew that he understood me on some level because I'd talk about the girls and he would cry. Tears would actually roll down his cheeks. Once I said, 'If you can hear me, squeeze my hand twice.' And he did it. He did it! But he would never do it for the doctors."

The doctors continued to treat Nick as best they knew how. He still exhibited no response to most painful stimuli; he reacted only to what the medical reports refer to as "deep pain" by "slight generalized flexion of his upper extremities." He was given a transfusion of nine units of blood when he developed a severe gastrointestinal hemorrhage. And because of continued spikes of high fever, likely due to a series of minor infections, his body temperature was monitored closely around the clock. He developed aspiration pneumonia at one point.

One question would surface again and again in the days, weeks, and months following the accident: Why had it taken so long to get Nick into the hands of experts who could help him? The Worthington hospital simply had not possessed the expertise or the facilities necessary to deal with the effects of rapid decompression. Kent Morstad, the emergency medical technician who had escorted Nick to the hospital, recalls wondering at the time why they hadn't pressurized Nick inside the suit and airlifted him directly to Minneapolis. The project had been at least nominally prepared for just such a medical emergency and Dr. Thompson had alerted the staff at Hennepin General more than a week before the launch. The phone number of the hospital had been included in the detailed flight plan. Somehow, in the confusion and rush to administer first aid, it seems that established emergency procedures had been second-guessed or even ignored.

But whether a different course of action would have altered the final outcome for Nick is doubtful. The long descent under the cargo chute following the decompression made any meaningful recovery the longest of long shots.

By May 3 the prognosis had apparently improved again. Nick's state was known as *coma vigil*, meaning that his eyes were open and roving. A standard diagnostic procedure involved flashing bright lights in the patient's field of vision and monitoring a variety of potential physiological responses. The doctors believed they were seeing some evidence of cortical brain activity. Dr. Hitchcock was quoted by the Associated Press as saying that the patient could still conceivably make a full recovery: "We feel that the neurological examination shows the possibility for recovery. While he is still in a coma, he is showing signs that he could recover. It's just a matter of waiting. He may never recover, or he may recover fully. At least there is still a reason for hope, and I guess his youth and vigor would tend to give him an advantage." Everyone looked for reasons to believe.

Even some of the reporters who had behaved so disgracefully in Worthington were cheerleading for Nick in their newspapers. Back in Sioux Falls, for example, the *Argus-Leader* ran a prominent tribute on its editorial page titled "Nick's Bold Flight": "Though his project terminated in failure, there is an admiration for the bold spirit that dictated the effort he made."

The most hopeful moment for Janice came on Nick's third day at Hennepin General. She was, as usual, sitting determinedly at Nick's bedside, pressing his hands between her own, trying to talk him—will him—back into consciousness.

"Nick, you can do it," she said. It was a mantra she repeated endlessly. "You can do it. All you have to do is try, and I'll help you." And then, according to Janice, his grip suddenly tightened on hers and he began to pull, as if he were attempting to pull himself up into a sitting position. "You're there!" she shouted, barely able to contain herself. "Nick, you're there. Just come over. Come on over! You can do it!"

Janice was weeping openly now, overwhelmed by emotion. And then, as suddenly as it had begun, Nick relaxed his grip and let go, as if exhausted, and settled silently back into the bed. After that, the doctors were never again able to detect brain activity. It was devastating.

The next day at breakfast, Janice came across an article in the Minneapolis paper quoting Dr. Hitchcock as saying that Nick had only a "one-in-a-million" chance to pull through. Furious, she marched into the doctor's office and all but threw the paper on his desk. "What do you mean?" she asked. "What does this mean?"

She recalls that Hitchcock sat her down and talked to her for what seemed like hours, explaining patiently that while there would be good days and there would be bad days, there really wasn't a great deal of hope that Nick would ever fully recover. Hitchcock told her that Nick's eye movement shouldn't fool her, that it meant nothing. "It's just his eye muscles working," he said. "There's nothing behind it. What we call the 'gray matter' of the brain—that's not working." He counseled her to try to accept the fact that her life with Nick as she'd known it was over. It would be excruciating, but she was going to have to find the strength to continue without the husband she had known.

Following the hope she'd felt only the day before when Nick had squeezed her hands and tried to pull himself up, this was the lowest moment for Janice since the accident. The reality of her situation began to sink in. "I thought I would die of a broken heart," she says. "It's hard to believe a person could feel so much pain and not be just consumed by it."

The real saving grace, once again, came in the form of Bill Jolly and Pioneer Parachutes. Jolly, at Pioneer's expense, reserved rooms for himself and for Janice at the Downtowner Hotel just up the street from the hospital and escorted her to and from Nick's room several times a day. It was an admirable display of professional and personal loyalty. Once, on their walk to the hospital on a beautiful summerlike day, Jolly asked Janice what she thought was going to be the hardest part for her.

It took her a few moments to respond. "You'd have expected me to say, 'Raising the children,' or something like that. But what I said was: 'That I never did get any roses.' So stupid, and yet for me at that moment, it seemed bigger than raising the children alone."

On the third day of Nick's stay at Hennepin General, Catherine Piantanida arrived in Minneapolis and was able to relieve Jolly and provide some support for her daughter-in-law. There were gentle suggestions that Janice should arrange a trip back home to see her daughters, but Janice would not discuss leaving Nick. She was still nearly oblivious to the outside world. One afternoon while she was making her way to the hospital from the hotel, a tornado struck Minneapolis. A policeman had to physically pull her off the street and drag her into a doorway as she insisted that if she was going to die she wanted it to be with her husband.

On each of the forty-eight days Nick Piantanida was a patient at Hennepin County General Hospital, Janice was with him for every minute allowed under hospital visitation rules. "She's here all the time, every day," a hospital spokesman told the local papers.

"One minute I'm optimistic," Janice told a reporter for the *Asbury Park Evening Press* who managed to get her on the phone one afternoon, "and the next I'm terribly frightened for him. I'm sort of in a haze, and sometimes I can't remember something I did a minute ago."

One question that Janice found difficult to answer kept coming up: Why hadn't she listened to the family and done more to try to stop Nick from continuing with such a dangerous activity? "I probably have enough influence to have stopped him," she replied. "But then I would have killed him on the inside. He's a very determined and a very stubborn man, and he really wanted this."

With Catherine in Minneapolis with Nick and Janice, and with Nick's prognosis seemingly up in the air, Florenz Piantanida was left to fend off the New Jersey press on his own. He chose not to share his criticisms of Strato-Jump with strangers, attempting instead to offer an explanation of Nick's motives that reveal not only hope, but a measure of paternal pride.

"Ever since he was a little kid," Florenz said, "he liked adventure. He wanted to dare things, to do them just because they were there. He was always doing things that were dangerous. I remember once when he was 13, he and a group of boys tied a rope to a tree and swung out over a steep cliff in Hoboken. He did many things like that. Now we just hope everything comes out okay."

Meanwhile, Janice's parents in Toms River, who were taking care of the two older Piantanida daughters, were also sought out for comment. "We have not actually sat down and told them [Donna and Diane]," Mrs. McDowell admitted to a local reporter, "because we don't want them to have to go through too much at once. We leave the radio on and talk about it in front of them, though."

Lee Guilfoyle at the Lakewood Parachute Center where it had all begun for Nick said that he also left the center's radio on all day in hopes of picking up news from Minneapolis. Most all of the jumpers there knew Nick, and even those that didn't were pulling for him. "That poor guy has had unbelievably bad luck," Guilfoyle said. "It's a damn shame."

Janice was grateful for the few visitors who came to see Nick in the hospital, including the Stefans and Don and Jeanette Piccard. But the two things that helped her the most during this ordeal were prayer (she and Catherine went daily to a Minneapolis church) and the cards, letters, and telegrams that came to her by the hundreds. There were notes from family and friends, as well as an amazing array of get-well cards from well-wishers around the world. Cards addressed to "Mr. and Mrs. Bonita" and "Mrs. Peontato" and one addressed simply to "Parachutist, Hospital, Minnesota" found their way to Hennepin General. Radio station KICD in Spencer, Iowa, had been broadcasting regular updates on Nick's condition, and listeners across the region from towns like Iroquois, South Dakota, and Cylinder, Iowa, had responded by mailing Janice religious pamphlets, holy cards, St. Christopher medals, and tiny Bibles. Some wrote letters recommending cures and therapies. Parachute clubs across the United States sent their condolences, as did a group of two dozen private pilots from the Estherville Airport.

Both Dick Kelly's wife, Kari, and Charmian Yost in Sioux Falls wrote regularly. Stanley Mohler of the FAA wrote from Washington, D.C., and the mayor of Brick Town, New Jersey, sent a huge bouquet of flowers. The president of South Dakota State University and Maj. Gen. Brooke Allen at the NAA both wrote, and esteemed Brig. Gen. J. W. Stillwell offered Janice any assistance the Green Berets might be able to provide. Sen. Harrison Williams, in his newsletter to constituents, included a tribute to Nick titled "A Brave Man." A receptionist from Williams's office sent a personal letter on congressional stationery and the entire congregation of a Wilmington, Delaware, church sent a card in which each member suggested a different Bible verse for Janice. A Tom Piantanida from Florida who had never met Nick wrote to say that he always "stood a little taller" when asked whether he was related to Nick. "Of all the members of the Piantanida family," he wrote, "only Nick will be recognized by history."

The fourth-grade class from the Douglas School in Minneapolis, a class that included Karl and Lucy Stefan's daughter, sent a bulging envelope full of handmade cards to Nick and Janice shortly after they arrived in Minneapolis. Like many other classrooms across the country, the students had been following the exploits of Project Strato-Jump since the first flight the previous October.

A friend and teammate of Nick's from St. Mary of the Plains College, Jim Moriarty, had become a basketball coach and math teacher at St. Pius X High School in Kansas City where for five years he had been regularly regaling his students with stories about the man he called "Tricky Nicky." One of Moriarty's students wrote to Nick on behalf of the class: "Coach Moriarty has always told us to watch the papers to see if you are doing anything. Since you have been making these jumps, all the newspaper clippings of you are cut out and read to all his math classes. Today at the beginning of class, Coach said we were to say a prayer for a special intention. When he asked us for our prayers, I was very surprised. The only other time that he has done this was when a member of our faculty died at Christmas time. Even last year when his oldest son was sick in the hospital, he never asked for our prayers."

May 8 was Nick's seventh straight day in a coma. But the doctors—continuing, in spite of Hitchcock's pessimistic private advice to Janice, to send mixed signals—were saying of the coma that it was "possibly a little lighter than it had been." Janice remained convinced that Nick could hear her when she spoke to him.

The problem with these kinds of comas, the doctors had explained repeatedly, is their unpredictability. Some patients take a turn for the worse and then recover suddenly and completely. Others linger unchanged for years. The hopeful but uncertain words from the hospital staff reflected the nature of medical science's relatively limited knowledge of such brain injuries at that time.

Back on the East Coast Vern Piantanida took matters into his own hands and contacted Dr. Roger Cracco, a nationally recognized neurologist and childhood friend who had grown up around the corner from the Piantanidas in Union City; his father was the Piantanida family doctor. Vern asked him to go to Minneapolis, consult with the doctors there, evaluate Nick's condition first-hand, and report back to the family. When Cracco returned, his news for Vern and the Piantanida parents—a verdict that the Minneapolis medical team had been unwilling to state publicly—was as straightforward as it was brutal: "There is no hope of recovery."

Meanwhile, Nick's friends and neighbors in Brick Town began organizing efforts to provide some financial assistance for Janice and the Piantanida daughters. A thrift-shop owner pledged that a percentage of her profits would go directly to the family and a letter to the *Ocean County Daily Observer* announced the establishment of a Nick Piantanida Fund. A senior-citizen's musical group called the Point Pleasant Poinsettias prepared a special program for the purpose of collecting donations to the fund, the Marge Fecher School of Dance presented a fundraising organ and dance recital, and the local Key Club threw a benefit party for the same purpose. An account was opened at the Pinelands State Bank to handle the proceeds, and the fund was administered by Mrs. James Hannon, the Piantanidas' neighbor on Nottingham Drive who had been taking care of the youngest girl, Debbie, ever since Nick and Janice had left for Sioux Falls back in April.

On June 15 the Plaza Luncheonette in Brick Town began selling toy parachutes for fifty cents a piece to raise money for the fund. They sold over one hundred on the first day. Altogether, it was a generous outpouring of affection, and in time Janice would be capable of appreciating it. But now all she could do was clutch Nick's all-but-lifeless hands and pray.

Bethesda

In spite of the doctors' earlier pronouncements that had often seemed to provide at least a window of hope, and in spite of Janice's faith and loyal vigilance, Nick's condition continued unabated (in fact, he had developed a duodenal ulcer along with his pneumonia), and Dr. Hitchcock's statements to the press became increasingly pessimistic. The outlook for Nick, he was now admitting, was "grave," and any significant recovery was "very improbable." Since there was little more that doctors could do for Nick, Hitchcock began to anticipate the next phase of his care—and the outlook wasn't encouraging.

"It's a great catastrophe," he said, "but there are people like him in nursing homes all over America. It may cost as much as $20,000 a year to take care of him."

On June 18, seven weeks after the accident, Nick was transferred by military transport—at the Department of Defense's expense—to the clinical center of the National Institute of Health (NIH) in Bethesda, Maryland, for "observation and supportive therapy." At Bethesda Nick was placed under the care of Dr. Maitland Baldwin. Doctors and medical researchers were understandably anxious to study the man who had survived near-explosive decompression at 57,000 feet, and Baldwin had been on the leading edge of—beyond, some who considered him a modern Dr. Frankenstein would say—neurological science. Some of Baldwin's research, supposedly under CIA supervision, involved bombarding the brains of lobotomized monkeys with radio waves. In one notorious experiment, he had reportedly attempted to transplant the brain of one monkey into the skull of another.

Remarkably, given the hundreds of astronauts and rocket-plane pilots who have journeyed into space, not to mention the balloonists who have gone to the very limits of the atmosphere, Nick's remains one of the very few cases of sudden decompression above the tropopause. In fact, with the exception of Peter Dolgov's accident in 1962, the only incident to rival—indeed, to eclipse—the disaster on Strato-Jump III occurred in the Soviet Union five years later. *Soyuz 11* was launched on June 6, 1971, and delivered three cosmonauts to the new *Salyut* space station. *Salyut*, the world's first orbital space laboratory, would represent a redemption for the Soviets after forfeiting the race to the moon. Cosmonauts G. T. Dobrovolsky, V. N. Volkov, and V. I. Patsayev spent three weeks on *Salyut* and set off on their return to Earth on June 29.

During the descent, at an altitude of about 100 miles, pyrocharges designed to separate the descent module from the orbital module malfunctioned, resulting in a pressure-equalization valve opening prematurely and depressurizing the capsule. The cosmonauts were not wearing pressure suits at the time and had only moments to struggle before succumbing. In spite of the exotic symptoms, rapid decompression actually kills in two quite prosaic ways: a sudden and total lack of oxygen starves the brain, and the development of air embolisms occurs when air passages begin to close. Recovery teams reached the three Russian bodies shortly after touchdown. All three were subsequently found to have blood in the lungs, massive hemorrhaging in the brain, high levels of nitrogen in the blood, and tissue damage from blood "boiling"—all reliable indicators of rapid death by decompression.

As a result of the *Soyuz* tragedy, subsequent Soviet space crews would wisely adopt the precaution—a standard feature of the United States' Apollo program by that time—of wearing full-pressure garments during the launch, docking, and re-entry phases of space flight.

It was all she could do just to hang on. Rather than return home, Janice elected to accompany Nick from Minneapolis to Bethesda. She had lost a lot of weight and was down to about ninety pounds. Without money or resources of any kind, and by this point without even much hope, she was compelled to accept assistance from the Red Cross who, through a local church, arranged for her to stay with a family in the area. Red Cross representatives were able to get her a cafeteria pass at the hospital so she could get her meals. Eventually, they also provided some much-needed counseling.

The NIH's policy only allowed visitors for five-minute intervals twice a day. So once in the morning and once in the evening Janice would make her way to the thirteenth floor where the patients with brain injuries were quartered. She saw men and women recovering from various types of specialized brain surgeries, an environment that she found both grotesque and fascinating. Dr. Baldwin would invite her into his laboratory-office to view the brain specimens arrayed in racks of jars: cats and chimpanzees and the deformed offspring of human incest. He explained the process of some of the blood chemistry analysis and other investigative work he intended to do with Nick. He was very kind to her and spent a significant amount of time explaining his research, why he was interested in Nick's case, and what he believed might be occurring inside Nick's brain. More important, perhaps, he convinced Janice that her husband was now—as the subject of potentially groundbreaking medical research—making a real contribution to humankind that was every bit as important as the one he had hoped to make with Project Strato-Jump.

As the weeks wore on, Janice's maternal instincts returned and she began to feel the absence of her children weighing on her. She knew she should be with them, but in spite of the guilt, she couldn't rationalize turning away from Nick. To leave him would be to surrender.

"They wouldn't let me see a lot of Nick," she says. "If they would have let me, I would have sat right in the room with him and been a martyr. I just felt like Nick didn't have anybody. I had to stay by his side. I was all he had. I couldn't abandon him." Nick's parents and brother would sometimes visit on weekends, but in her grief and despair Janice barely noticed.

In spite of the fact that Nick's neurological condition still showed no signs of improvement, Red Cross counselors finally intervened and convinced Janice that for her own sake she needed to go home to New Jersey and spend time with her daughters. They worked out a schedule that would send her home for three days and bring her back to Bethesda for three days. Through this regimen, Janice gradually reacquainted herself with her family and got comfortable with a new routine, learning that her constant presence was not in fact required and that Nick's condition was beyond any help within her power to provide.

Dr. Baldwin had promised Janice that when it became clear Nick was nearing the end, he would arrange to have the patient moved, at government expense, to a facility closer to home to avoid saddling Janice with the cost of transporting the body from Bethesda. So on August 25, 1966, ten days after his thirty-fourth birthday, Nicholas J. Piantanida, still comatose and feverish with a body temperature of 102 degrees Fahrenheit, was transferred to the Veteran's Administration Medical Center in Philadelphia. Janice continued to maintain a vestige of hope, telling a local reporter that Nick would continue to receive intensive care at the VA hospital and that he would immediately be transferred back to Bethesda the moment his condition began to change. "He has shown some improvement," she insisted, "but not nearly enough. I just try to keep my chin up and hope and pray."

But even Janice seemed to know that the time had come. Four days after being moved to Philadelphia, Nick died suddenly on August 29. Dr. Baldwin phoned Vern seeking the family's support for an autopsy, saying that Nick would have wanted it for science's sake. Janice gave her permission for the autopsy to be performed but asked not to be informed of the findings.

The official cause of death was listed as "massive encephalomalacia due to hypoxia with necrosis of the cortex and multiple infarcts of the basal ganglia." A layman's translation: softening of the brain due to oxygen deficiency, with tissue death in the nerve centers that handle the interpretation of sensory impressions. A long list of secondary ailments was assumed to be a collective side effect of the brain damage and the four months of comatose inactivity. Nick weighed barely 150 pounds at the time of death.

Return to Union City

Janice's Brick Town neighbors drove her to the VA hospital in Philadelphia. Vern and Catherine Piantanida drove down from Union City. In spite of having seen Nick slowly wither over the weeks since May 1, confronting his lifeless body was a terrible shock for Janice. She began to wail uncontrollably and nurses took her to a private room where she cried for what seemed to her like hours.

Funeral arrangements were tricky. Janice preferred a small, simple ceremony in Brick Town, and said so, but Catherine was adamant that Nick should have a High Mass and a full Catholic funeral service in Union City. Even in her disjointed state, Janice was sufficiently cognizant to realize that with Nick's death, her influence over family matters had evaporated and that Nick's parents and brother—whose longstanding concerns and fears about Nick's high-altitude activities had been vindicated—were now in control. So a compromise was reached: the body would be laid out for a few hours in Brick Town on Thursday evening and again on Friday morning, and then moved up to Union City for the funeral on Saturday of Labor Day Weekend.

Vern, Catherine, and Florenz went down to Brick Town to be with Janice that night. Vern, who had seemed to Janice like anything but an ally through the Strato-Jump years, now attempted to console her.

"You okay?" he asked.

"Yeah," Janice answered. "I think I'm fine."

But when she got to the little funeral parlor and saw Nick in the casket, she lost her bearings once again and crumpled to the floor. She was given a sedative and put to sleep in her neighbor's house for the night.

The next afternoon, everyone made their way to Union City. A doctor had given Janice a strong tranquilizer to help her through the ordeal, and her father, James McDowell, had come along to accompany her through the day. When they arrived at the church on Saturday morning, several hundred people had already lined up to attend the service. To Janice it was a circus atmosphere reminiscent of the Kennedy funeral three years earlier. In fact, it was the kind of hero's farewell that Nick might have envisioned for himself.

The wake lasted for two days, and crowds packed the funeral home for the afternoon and evening visiting hours both days. Janice was escorted into the viewing area and was then led through a string of rooms crammed full of hundreds of dozens of flowers where, in one hallway, she came across a dramatic arrangement of one-hundred long-stemmed red roses, courtesy of Bill Jolly who hadn't forgotten Janice's comment in Minneapolis. "Just seeing those flowers from Bill helped me get through that day," Janice recalls. "As if he wanted to make sure I had at least one rose for each year of the rest of my life."

Most people thought Janice looked drugged, which of course she was. Well-wishers approached warily and said things like "Oh, Nicky looks so beautiful!" Janice could only stare blankly from within her grief-stricken haze and think, So *beautiful*? What are you talking about? He looks so *dead*.

The funeral was almost as hard on Janice's father, who had developed a great affection for Nick and who was himself devastated by the turn of events. When Mass was over, James told Janice firmly that it was now time for her to take her children and reclaim her life. But James himself was never able to recover from his son-in-law's death. Shortly after the funeral, he slipped into a deep depression. James McDowell died a few months later. "He died of a broken heart," Janice says. "He cried every day. There's no other way to explain it. He really loved Nick. It just . . . broke his heart." Vern had never seen his own father cry before, but Florenz was so crushed by the death of his first son that he wept openly for extended periods of time and refused to acknowledge or to speak of the accident for nearly a year following the funeral.

Nick's magnetic personal qualities as well as his lifelong instinct for self-promotion resulted in an outpouring of posthumous affection. Janice received more than two thousand letters from around the world. Bill Jolly, of course, attended the funeral in Union City. Ed and Charmian Yost, along with Dr. Frisby, came from Sioux Falls to pay their respects. Roger Vaughan took time off from another *Life* assignment to be there. Vaughan was horrified to see the emaciated body of his friend displayed in an open casket.

Nick Piantanida was buried in Holy Cross Cemetery in North Arlington, New Jersey. The inscription that Janice asked to have engraved on the stone marker read: I AM AN ACME OF THINGS ACCOMPLISH'D, AND I AN ENCLOSER OF THINGS TO BE. Almost thirty years later, Florenz and Catherine would die within a year of each other, and both would be interred in the same plot with Nick.

Barry Mahon and team edited their Strato-Jump footage into a stirring tribute to Nick called *The Angry Sky*, which was shown at the Surf Club in Lavalette on the Jersey shore as a fundraiser for Janice one Wednesday night that fall. Phil Chiocchio, who shot some of the footage and who was listed as assistant director on the film, was still underage at eighteen and tried to sneak in with a friend's expired boat-operator's ID. He was thrown out by the club's bouncer but was later allowed in just long enough for the screening.

Phil remembers an emotional crowd of Nick's friends and colleagues sitting stunned and mute at the film's conclusion.

The film includes some rare early free-fall footage shot with a helmet-mounted camera by Lee Guilfoyle at Lakewood, and a fairly corny reenactment of Nick's first parachute jump. It also includes footage of all three Strato-Jump launches. The most gripping moments occur late in the film as Dr. Thompson and others struggle to administer first-aid following the Strato-Jump III touchdown in the Minnesota field. There is no record of precisely how much money was collected at the Surf Club, but it was enough to allow Janice to pay some back rent and settle a number of overdue bills. There was even enough to buy her a car, even though she'd never learned to drive. Vern showed up in Brick Town one sunny morning, piled the kids in the back, and told Janice to get behind the wheel. He instructed her in the basics and then directed her to the busy New Jersey Parkway and into the center lane.

"If you can drive here," he told her, "you can drive anywhere in the world."

Roger Vaughan filed his story with his *Life* editors only days after Nick's tragic finale. The account of a triumphant Project Strato-Jump had been planned as a cover story, but the unfortunate turn of events would move it inside the magazine in favor of a fashion piece on the latest sartorial habits of the U.S. male. When the article appeared in the May 13 issue, accented by Robert Kelley's stunning photography, Vaughan was furious but hardly surprised. One of the magazine's missions at that time had been to locate and promote stories of American heroes, and Vaughan was livid at what the editorial staff on the thirty-fourth floor of the Time/Life Building had done. They'd taken his copy and transformed it into saccharine melodrama, casting Nick as a shining knight brought to ground by an overwhelming foe. They even invented a sentimental incident to augment the story of Nick's childhood illness (which even Vern Piantanida remembers as "no big deal"):

> But when a doctor diagnosed osteomyelitis . . . and put Nick on crutches, he devised his own cure. What he did was to hide the crutches in an alley one day on his way to school and thereafter, step by painful step, repair his legs by a sheer act of will.

That's not exactly how it happened and that's not the way Vaughan had written it. He'd presented Nick as an admirable—but flawed—man. "I thought he was a good guy," Vaughan says. "I thought he was a *real* American, in the pioneering spirit. But he was flawed. He's a guy who would always frighten you just a little bit."

In the piece he had submitted, Vaughan had spelled out the prevailing opinion on what had occurred in the gondola at 57,000 feet. It was the very crux of the difficult story he had been compelled to write. The editors had simply deleted Vaughan's explanation for the disaster and supplied their own safely ambiguous version: "his suit suddenly—and inexplicably—lost pressure."

"I was pretty upset about it," Vaughan remembers. "A detestable bunch. All they wanted to do was sell washing machines. If there was anything in any story that wasn't going to sell washing machines, it was going to go."

In truth, the issue of *Life* in which Vaughan's story appeared carried no advertisements for washing machines, but most every other appliance necessary for mid-1960s upper-middle-class domestic bliss was represented, including a reclining chair called—in an example of the hold high-altitude aviation and space travel were beginning to exert on the American popular imagination—the Stratolounger: "Your cares float away!"

As the widowed mother of three, Janice's cares were very real and solidly earthbound. Not long after the funeral, she packed up the girls and drove them to Pattersonville in upstate New York, not far from Schenectady, where her mother's family had owned farm land for generations. Not much later, she used what was left of her insurance settlement, the remaining funds from *The Angry Sky* benefit, and the Brick Town fund to start construction on a modest house. Using the interest earned by the money from the girls' insurance proceeds, along with Social Security benefits, Janice was able to start a new life for her shattered family. She broke off contact with most of the people she and Nick had known in Union City and Brick Town and elsewhere. "I sort of dropped out of sight," she recalls. "I don't think people had a way of getting hold of me. And I wanted it that way."

Janice remarried, but the fallout from Nick's death was like a shadow that never receded. Eventually, in spite of the birth of Janice's fourth child, a son named Brian, the burden of past events became too much and the marriage imploded. Her second husband, Janice says, just couldn't take it. But she never blamed him. "It was too much for him," she believes. "It would be too much for any man, I think."

The one contact Janice kept up from the Strato-Jump days was with

JADODIDE (*clockwise from left:* Diane, Janice, Donna, Debbie). (Courtesy Janice Post)

Charmian Yost. Charmian, she thinks, understood better than most because her own husband Ed was so much like Nick.

"It was just hard to believe that Nick could die," Janice says. "It was really just hard for me to believe it, even years later. It was like it didn't make sense, you know? I mean, he was like Superman! How could he die?"

The wedding ring that she had tucked into a pocket of Nick's pressure suit prior to the final flight has never been recovered.

7 "MAGNIFICENT FAILURE"

"E*mer*gen . . . !"

It had been an excruciating drama for the crew monitoring voice communication at Joe Foss Field that morning. Most of them heard the distinct and sudden hiss of air, the "whoosh." They heard the urgent utterance that followed, described inaccurately by some as a "gasp" ("as though he had been hit in the stomach"), but which had clearly been an attempt by Nick to say something. And they heard Nick's final transmission cut off in mid-word. About these events—at least among those who were on the scene in Sioux Falls that day—there is little dispute. Some accounts over the years have mentioned "anguished grunts" and "screams" that were supposedly transmitted from the gondola. The following description in the August 2001 issue of *Wired* is typical of such reports: "He was still on his way up at 57,600 feet when ground control staff heard a scream and then a monstrous gush of air come through their monitors." Such descriptions have long been discredited by witnesses and by tape-recorded documentation of the flight's ground-air communications. But what occurred inside the gondola immediately prior to the "whoosh" has been the subject of controversy and confusion ever since.

When reporters cornered Ed Yost at Worthington Municipal in the agonizing hours following the incident, they asked him to speculate about causes. And Yost was willing, figuring his guesses were as good as anybody else's. He initially, and perhaps naturally, suspected a failure of some element of the survival system: the pressure suit or the helmet. Yost would later suggest that Nick might have inadvertently decoupled an oxygen line—which, if true, would have been the height of irony given the problem with the oxygen disconnect on Strato-Jump II. Marvin McCall, whose devotion to Nick had been obvious to everyone and whose very foundation had been shaken by the accident, had also spoken to the press at the hospital in Worthington, and his remarks echoed Yost's; McCall said that the suit had apparently failed. But to get beyond speculation, a full postflight investigation of the entire system—suit, helmet, oxygen supply—would be required.

"A Grave Injustice"

The David Clark Company had not one but two experienced technical representatives on the ground in Sioux Falls for the final Strato-Jump flight. Given that the pressure suit was a valuable and militarily sensitive piece of equipment, and given that it did not belong to Nick or SPACE, Inc. but was merely on loan to the project, a David Clark employee was assigned to escort it to the launch location and back to Worcester following each of the three flights. Dick Sears, who had been involved with Nick since his earliest visits to Worcester more than a year earlier, had been the onsite representative and observer assigned to Project Strato-Jump for each flight, and he was joined for the May Day attempt by technician Tom Smith. Between them, the two men had been involved in dozens of high-altitude flight programs. Like most of the Strato-Jump team, Sears and Smith had developed an affection for Nick. But it went deeper than that.

When asked by a writer for the *National Observer* why David Clark had decided to get involved with such a risky project, Sears responded, "Why are we doing it? I guess we have just developed a sneaky respect for this guy."

Sears and Smith had arrived in Sioux Falls on the evening of April 27 in time for a project team meeting at Raven Industries the following morning. They participated in the dry run of the launch procedure. Everything went well except for some minor problems with Bob Kelley's cameras and some confusion over the elements and sequence of the emergency procedure. Two days later Sears, Smith, and Marvin McCall inspected and tested the pressure suit and helmet, and repeated the entire dry run. Sears's report would say that the day's tests "went off exceptionally smooth." When all the tests were complete, the suit was dried, inspected, and stored.

On April 30 the suit was reinspected, retested, and re-stored. There were press conferences and more team meetings that day, the final meeting at the launch site at Joe Foss Field. In order to reduce the chances for confusion on launch morning and, one suspects, to keep curious reporters from getting in the way, it was agreed that the only individuals who would be granted access to the administration building in the National Guard hangar once the suit-up process had begun would be Nick, Janice, McCall, Sears, Bill Jolly, Bob Kelley, and Jacques Istel.

Early on the morning of May 1, Sears and McCall worked together to assist Nick in donning the Dave Clark Special. Just before Nick left the suit-up area, the visor on the helmet was closed and the entire system was pressurized. Everything checked out, so the suit was deflated. They then pressurized it one final time and repeated the test. All systems were operating perfectly and Nick was able to head out to the gondola with high confidence.

But as Nick climbed into *JADODIDE* that morning, Sears was not happy to see a number of individuals, some of whom he didn't know, such as members of Barry Mahon's film crew, the General Electric technicians, members of the press, and perhaps others, milling about the gondola. Unbeknownst to the David Clark field reps, Nick had personally ordered that access restrictions to the launch site and the gondola be greatly relaxed for that flight to accommodate the needs of the various parties involved. Sears grabbed McCall.

"Marv," Sears said, "you should be the last person in that gondola and do a final check before it lifts off."

McCall replied: "I already did it."

"But you should be the last man in there, Marv," Sears repeated, concerned that things were getting a bit out of hand in the excitement surrounding the launch.

McCall could only shrug.

When Nick reported the emergency at 57,000 feet less than an hour into the flight, Sears and Smith immediately jumped into Sears's rental car and sped out to the projected landing area. They followed the ambulance to Worthington Municipal like everyone else and, like everyone else, tried to ascertain what had happened. Sears was concerned to hear reporters saying that the suit or helmet had malfunctioned, probably as a result of some of the comments from Yost and McCall. It was exactly the kind of thing all of the survival-gear contractors—along with the Air Force and the Navy—had feared and been desperate to avoid. Sears overheard a reporter in one of the hospital's telephone booths phoning a radio station. The reporter was saying that the helmet visor had cracked and, curiously, that doctors were calling for blood donors.

"That's a damned lie!" Sears yelled. The reporter pulled the phone-booth door shut. When he finished the call and came out, Sears confronted him.

"Where did you get that garbage?" he wanted to know. His fury rising, Sears informed the reporter that there was no evidence that the visor had cracked and that there had most certainly been no call for blood.

The reporter's response: "Well, we can always retract it later."

David Clark technician Tom Smith was a young man in 1966 and was shocked at the methods and behavior of the press in Worthington. "Of course, we were very sensitive about our equipment," Smith recalls. "And we knew that there hadn't been any failure that we had seen. And to hear what they were reporting second- and third-hand. . . . I was right there with Dick when this one fellow was calling in to his editors in New York that a cracked visor was the cause of the accident. I was just crushed. So angry. They were so casual about it. It was really sensationalism at its worst."

The following morning, readers all across America would open their newspapers to claims like this one that appeared in the *Minneapolis Tribune:* "the visor on his helmet cracked and discharged his oxygen supply. . . . Engineers would not comment on what caused the visor to crack." An Associated Press story that was picked up by a number of papers that same day had a slightly different explanation: "He wore a pressurized flight suit and helmet, but the pressure was lost when a pneumatic seal around the helmet face plate apparently failed." This story attributed the failed-seal theory to Ed Yost but included a comment from Yost that "the suit showed no breaks and there were no cracks in the helmet." The UPI story that appeared in the *New York Times* the next morning quoted

Jacques Istel, misidentified as the president of Raven Industries, saying simply that "equipment trouble" had caused Nick to descend.

Newspaper reporters on deadline weren't the only offenders. Magazines and books have for years printed accounts of the accident stating unequivocally that equipment failure was to blame. Veteran parachutist and respected aviation writer Bud Sellick wrote about Piantanida in a book published a full five years after the accident on Strato-Jump III. "Apparently," Sellick wrote, "his face shield had blown out of the pressure suit." A historical overview of lighter-than-air flight published in 1977 offers yet another explanation for the accident: "a seal on his pressure suit gave way at fifty-seven thousand feet." Even the medical staff at the National Institute of Health repeated an ill-founded rumor in its clinical report on Nick's case: "the patient suffered a hypoxemic, hypobaric cerebral injury resulting from a free-fall in a gondola from 57,000 feet, during which the pressure system in the patient's pressure suit failed."

Sears and Smith finally caught up with Dr. Thompson later that morning and got a brief status report on Nick's condition. Moments later Sears, Smith, and Marvin McCall found Bill Jolly who had some new information. Jolly informed them that Raven personnel had already reviewed the voice tape from the flight, and that the sound immediately following the rush of air was definitely more than just a grunt. Jolly had been told that it sounded like Nick had said the word "visor." (This conclusion would later be retracted by Raven.) Raven personnel at the landing site had also told Jolly that they had found "pieces of the visor" at the site.

Sears was flabbergasted. "That's impossible," he said. "We have never lost a visor. But if the gondola doors were open and it *did* blow off, it would have gone in one piece and wouldn't be anywhere around the landing area."

Sears asked about the helmet itself. Jolly told him that it had been picked up by Raven employees and returned to the plant. Sears tracked down Ed Yost at the hospital and asked him to contact someone in Sioux Falls and demand that the helmet be locked up the minute it arrived so as to prevent its being tampered with.

Once the suit had been cut off of Nick in the emergency room, Sears and Smith stuffed the remnants of the Dave Clark Special's orange fabric into pillowcases and managed to sneak them out the rear of the hospital without being seen by the press. They drove immediately back to Sioux Falls, arriving at the Raven plant about the same time a truck carrying the gondola pulled in. They decided to keep the suit locked in the trunk of the rental car until the Raven labs could be cleared of news people.

Once inside, the first thing Sears did was to inspect the helmet. What he found was that the sunshade had in fact shattered, presumably from impact as Nick was tossed about the gondola on landing. A shattered sunshade was not an unprecedented occurrence. David Clark frequently lost sunshades during windblast tests. However, as Sears had expected, the pressure visor was completely intact and fully operational. He put the helmet on and ran a quick pressure test with a 75-psi oxygen line, finding nothing to indicate a failure in either the visor or the face seal. Ed Yost had returned to Raven by this point and agreed to issue a call to the media outlets announcing that, on the basis of preliminary tests, nothing indicated a malfunction of the suit or the helmet. Sears had rented a camera shortly after arriving in Sioux Falls in order to take photos of the launch. He used it now to document the condition and configuration of the helmet, as well as what was left of the sunshade and the suit. Once that task had been accomplished, Sears and Smith returned to the Town House for some much-needed sleep.

The next morning, with Nick by now receiving treatment in the hyperbaric chamber in Minneapolis, Sears phoned Jacques Istel's room and went over the results of the previous day's events. He asked Istel to follow up Yost's call to the press with one of his own. Istel agreed, and his informational calls to selected reporters included the statement that "a grave injustice" had been done to the David Clark Company by the hasty, fragmentary stories that had gone out over the wires the previous afternoon. He promised that a full investigation would be conducted and that the complete results would be made public later that same day.

Sears and Smith spent the rest of the morning in the Raven labs running an exhaustive systems check of the helmet, the suit controller, and the breathing regulator, tests that were conducted in the *JADODIDE* gondola using the onboard oxygen system. The results were consistent with those they had gotten the night before, and a new press release was typed up on Raven Industries letterhead. All of the involved Raven personnel, available Strato-Jump crew, and contractors were assembled and the release was read aloud. Once everyone had indicated satisfaction with the contents, the release was delivered to the local news media and to *Life* magazine.

The Raven press release read, in part, "Engineers and technical personnel involved in the preparation of this physiological protective equipment announced today that all elements of this system have been checked out and no evidence of malfunction has been found. A statement from the aeromedical doctor who first reached the stricken parachutist has not yet been received. It is hoped that he, or Piantanida himself, may be able to provide an explanation as to the cause of the accident."

Nick was not going to be able to contribute to that explanation, but before leaving Sioux Falls Dick Sears did get an opportunity to talk with Dr. Thompson, as well as with Raven pilots Heidelbauer and Tomnitz. These were the first three individuals to arrive at the scene of the landing. Dr. Thompson reported that "Nick's face was exposed," indicating that the helmet visor was in the up—or partially opened—position. The pilots reported disconnecting the oxygen hoses, both the helmet hose at the suit controller and the supply hose coming into the controller. It was also reported that the breathing supply from the gondola's oxygen tank to the suit (via the hose with the quick disconnect) was disconnected, though neither the pilots nor Thompson could recall disconnecting it. In addition, the gondola's oxygen tanks were nearly full, exhibiting normal use, but the emergency jump cylinders in Nick's backpack were empty. The ventilation unit, which ran on liquid oxygen, was still operational and venting the suit when the emergency personnel arrived on the scene.

The next morning, May 3, Tom Smith—on another assignment from David Clark—caught a plane for Sacramento. Sears stopped by the Raven plant and picked up a copy of the flight voice tape he had asked Ed Yost to provide, and caught a 10:00 A.M. flight for Boston with all of the gear his company had loaned Project Strato-Jump.

Testing of the suit and helmet continued at the David Clark plant for a full month following the flight of Strato-Jump III. A technician donned an operational A/P22S-2 suit along with Nick's helmet and the suit controller from the Dave Clark Special Nick had worn. Everything was repeatedly checked out to the altitude equivalent of 85,000 feet.

According to Dick Sears's final report: "All known information points to a loss of oxygen. At this time any explanation of the cause of the O2 loss is strictly conjecture. There appear to be three basic possibilities: 1) An inadvertent disconnect of an oxygen line. 2) Bad oxygen which could have had moisture in it which could freeze and block an orifice. 3) Accidental or intentional deflation of the visor seal." Sears's list is reasonable, but if an interruption of Nick's oxygen supply from the tank to the suit (either from a malfunction at the tank or a breach in the hose) had occurred, it would have been a fairly simple matter to have switched to the bailout supply. Nick knew the procedure for making that switch quite well. He had already performed the switch several times in decompression chambers and had been planning to do it again later that morning when he reached peak altitude.

The David Clark Company, whose reputation and valuable contracts with the U.S. Air Force and NASA were conceivably riding on the results of the investigation, issued yet another press release to announce its findings, this one signed by company president John Flagg. It made five emphatic, if not quite desperate, points:

1. The visor (facepiece, lens) did *not* crack.
2. The pneumatic seal around the edge of the visor did *not* rupture.
3. The breathing regulator has been tested and performs normally.
4. The suit pressure controller has been tested and performs normally.
5. The actual reason for the failure of the oxygen supply is still unknown.

Flagg included some additional comments: "We will continue to investigate the matter in an attempt to contribute to the solution of the cause of the accident. We do make the very best spacesuits in the world. We specialize in lifesaving personal equipment, and anything less is unthinkable."

Nick himself had often expressed his supreme confidence in David Clark's design and engineering. As he once told Janice: "With all I have to worry about, the one thing I never worry about is my suit."

But what about the oxygen and the oxygen system? Could Nick's oxygen supply have been contaminated?

The procedure for handling, inspecting, and testing the oxygen system prior to the flight was thoroughly documented in the project planning documents. On April 21, a full ten days before Strato-Jump III lifted off, all components of the breathing oxygen system arrived in Sioux Falls. On April 22 all fittings were checked and all oxygen cylinders were topped off. On April 27 the fittings were again checked and the cylinders were again topped off. Two days later, on Friday, everything was once again checked and topped off:

main onboard supply, reserve onboard supply, both bailout bottles, and both walkaround bottles. On Saturday the oxygen equipment was moved to Joe Foss Field, rechecked, and topped off. Earl Clifford, an experienced field representative for the Firewel Company assigned to Project Strato-Jump, was in charge of these activities.

On launch morning Clifford was scheduled to have inspected, checked, and topped off everything one final time by 3:45 in preparation for the suit-up operation. And while there is no documentation to establish the precise timing of events in the suit-up area, there is no reason to suspect that the oxygen system was anything less than safe and reliable. Postflight inspections found no traces of contamination in any of the oxygen supplies or any evidence that any of the fittings had iced up or otherwise been compromised.

The bottom line is that the oxygen had been inspected and checked repeatedly by experienced personnel prior to the flight. Bad oxygen seems like an unlikely explanation for the emergency in the stratosphere.

Janice, perhaps naturally, would always suspect some sort of cover-up, seeing Sears and Smith hustling the suit out the back door of the hospital, believing that the David Clark Company was in an unseemly hurry to abscond with evidence that might have helped explain her husband's accident.

"Nick was too careful," she says. "I believe it was the equipment that failed. I know it wasn't Nick." She has always assumed that the problem had been somewhere in the equipment and that the David Clark reps had something to hide. However, as Yost pointed out later, regardless of the condition of the suit: "As a contractor, that would be the first thing one would do is get your equipment out before somebody blames you." Sears and Smith may have worked too quickly to secure their gear from Janice's perspective but not quickly enough to escape public blame.

A UPI story, not published until August of 1966, offered the first real glimpse of the David Clark Company's true suspicions of the cause of the accident: "Inspections of Piantanida's space-style suit and helmet revealed no structural deficiencies, according to the equipment manufacturers. The cause of sudden depressurization, they said, was something that only Piantanida knew." This was as close as the company would ever get to publicly suggesting that the accident on Strato-Jump III had been Nick's fault. The David Clark employees who had worked with Nick had all liked him and had all been impressed by his physical condition, his enthusiasm, and his willingness to overcome his inexperience by preparing and training diligently. Nevertheless, in the end, the consensus in Worcester seems to have been that Nick was simply too impatient and allowed his judgment to be clouded by his own eagerness to succeed.

Only moments after the whoosh had come across the radio and after Nick had tried to say the word "emergency," moments after Yost had issued the cut-down order and Rupp had hit the pickle buttons, a number of those on the ground were sure they knew what had happened. Roger Vaughan remembers hearing somebody at Joe Foss Field that morning say it aloud: "Goddammit, he flipped his visor up again!"

Guessing Game

The helmet Nick wore on all three Strato-Jump flights, a shell of rigid fiberglass, was supplied with an outer tinted-Plexiglas sunshade and an inner clear-Plexiglas pressure visor. The sunshade is raised simply by guiding it up with the hand; once the shade is in the up position, the pilot must depress a spring-and-ratchet-type disc control on the right side of the helmet in order to lower the shade. This effectively prevents an opened shade from snapping shut with sudden movement.

The movable pressure visor is coated with an electrically conductive gold film that provides resistance heat to the visor to clear condensation and at the same time creates some warmth for the pilot's face. With the visor in the down (or closed) position, a pneumatic pressure seal automatically inflates with 70-psi oxygen to seal and secure the visor to the helmet shell, creating an airtight zone around the face—referred to as the "oral/nasal breathing cavity"—that is automatically maintained at a slightly positive pressure with respect to the rest of the suit. This sealed facial compartment is supplied with 100 percent breathing oxygen, unlike the rest of the suit, which also carries ventilation gases and the pilot's exhaled gases. With the visor seal pressurized, the only way to open the visor is to depress another spring-and-ratchet-style disc on the left side of the helmet (known as the "left-hand bearing") that deflates the seal. If the wearer depresses this bearing while the suit is pressurized, it will dump pressure not only from the facial compartment but from the entire suit system as the gas from the suit would be expelled past the faceseal and rush out through the unsealed visor opening.

Upon release of the left-hand bearing with the visor in the down position, the helmet seal is designed to reinflate immediately, resealing and securing the pressure visor. If a pilot depresses the left-hand bearing and opens the visor, which deflates the seal, he must manually raise the visor to the full up (or open) position before closing it again. In other words, once the pneumatic seal is manually deflated and the visor is open, the wearer must depress the left-hand bearing a second time before the visor can be closed and the system resealed.

It was a common belief among the Strato-Jump crew and others on the scene that Nick had a bad habit of opening the visor both before and after suit pressurization (in other words, both below and above 35,000 feet). There were reports that he had been cautioned repeatedly by the pressure-suit experts—presumably Marvin McCall and others present for his decompression chamber runs—not to touch the visor after the system had pressurized. When the accident occurred, several of the ground crew immediately leapt to the conclusion that Nick had in fact opened the visor. And that's precisely what the story Roger Vaughan originally filed with *Life* magazine said.

"I came away from it without much question," Vaughan recalls. "According to the guys, he was in the habit of doing it—despite them telling him not to. Why did he lift his visor? That was really the only question. And I've never heard any different."

But the *Life* editors, understandably interested in avoiding any messy legal entanglements, preferred not to publish speculation and stuck with an idealized portrayal of Nick Piantanida as a by-the-book American hero who could do no wrong, rejecting the courageous but flawed adventurer of Vaughan's portrait.

If the speculation had merit and Nick *had* opened the visor, a number of possible explanations for such behavior must be considered. One theory holds that Nick briefly opened the visor in order to clear condensation from the interior of the visor. Visor condensation had been a serious problem for the pilots of some of the Air Force's manned scientific balloon flights several years earlier. In fact, a fogged-up visor, effectively rendering the pilot temporarily blind, was one of a high-altitude balloonist's greatest fears. Joe Kittinger had experienced it—and been quite frustrated by it—on his first Excelsior ascent. Darrel Rupp recalls some of the conversation following Nick's final flight: "I remember the speculation that he'd opened his facemask to get moisture out of it."

In the account of Project Strato-Jump presented in his 1980 book *Parachuting Folklore: The Evolution of Freefall*, master parachutist Michael Horan describes the condensation theory: "Others speculated his face mask had become fogged while passing through the troposphere, thus preventing him from seeing beyond the gondola. On numerous occasions, Nick experienced similar difficulties, and lifted his face mask visor and closed it rapidly to clear off the goggles. Perhaps this time the visor didn't snap shut so securely, and Nick was robbed of precious air escaping from his pressure suit."

The David Clark helmet Nick wore was a much improved piece of hardware when compared to the helmets the Manhigh and Excelsior programs had used. Not only did Nick's visor contain the superior electrical gold-film defroster/heater, but a spray bar on the helmet's interior directed the incoming flow of oxygen across the inside of the visor in an effort to clear any moisture. Nick did, of course, experience the annoying phenomenon of the spray bar blowing his facial perspiration onto the visor during the ascent on the both the second and third flights. On Strato-Jump II it had occasioned the strange but lighthearted exchange with Marvin McCall about opening the visor to clear the moisture. Yet in spite of Nick's comment that he had a "puddle" in the helmet, the visor heat seemed to have dried things out in short enough order. Nick did not mention the problem again and neglected to even refer to it in either his oral or written postflight reports of the second mission. While Nick remarked on visor condensation twice in the course of the third flight, he also mentioned that he was able to clear it effectively by activating the visor heating mechanism.

Raven pilot Al Tomnitz has always believed that Nick opened the visor, but he doesn't believe it had anything to do with condensation:

> He [Nick] commented that the sun was so bright. And you know he had that pressure helmet on, and one knob on one side would let the [outer sunshade] visor down and one knob on the other side would let the clear [inner] visor go up. I think he pushed the wrong one. Instead of letting the dark one come

> down and shield his eyes from the bright sun, he pushed the wrong one and it depressurized him.

Others have speculated that Nick's nose itched and that he opened the visor in order to scratch it. Another theory holds that Nick had an odd but long-standing habit of banging the side of his helmet with the heel of his hand, and that he accidentally smacked the visor bearing and deflated the suit. A close friend of Dan Quinn, the Lakewood parachutist who knew Nick well and who assisted with the chute-up procedure prior to the May Day flight, recalls this line of thinking: "According to Dan, he [Nick] had a very bad habit of striking the side of his helmet. I remember him telling me that when they did their practice jumps, Nick would always bang the side of his head. Dan's theory, and I think Barry Mahon agreed with him, was that when he was up there, he banged his helmet and there was a button there or something that released the pressure. I don't know whether it was an ear itch or what it was. He was warned and warned and warned, but he kept doing it."

An article published in *Parachutist* magazine in 1978 claimed that Nick, "using a trick which had worked before, tried to squeeze his temples to make the pain [in his left knee] go away" and inadvertently "clicked open the switch controlling the pressure in his helmet." It's difficult to imagine how Nick could squeeze his temples through a hard-shell helmet.

One morbid line of reasoning suggests that the project was laboring under such a mountain of debt that a distraught Nick, with no hope of ever repaying his creditors, committed suicide aboard Strato-Jump III by opening his visor. Those who knew Nick, or even those who were around him for any length of time, dismiss any such suggestion out of hand. As Janice said, "Nobody on this *earth* ever wanted to live more than Nick Piantanida."

The project planning document for a later, stillborn high-altitude jump program called "New Excelsior" captured the consensus of most of those who've seriously studied Strato-Jump: "On May 1, 1966, while ascending through 57,000 feet, he somehow partially opened his helmet visor, causing de-pressurization . . . resulting in his death." But the report goes further: "A later Air Force investigation determined that there were no problems with the equipment, and the opening of the visor had to be a conscious action on the part of Mr. Piantanida. Why he took this action has been a matter of speculation to this day." In fact, there is no evidence that the Air Force ever conducted a formal investigation into Project Strato-Jump.

Disaster in Waiting

Dick Sears's postaccident report to the David Clark Company, a thorough but properly cautious document, included an interesting note:

> Ed Yost said that Nick told him that at about 50,000 ft. he gets a skin tingling sensation and one of the things he can do to correct it is to depress the left hand bearing knob deflating the visor seal, thereby deflating the suit, let it go and immediately reinflate the suit.

Nick had apparently complained of being "overpressurized" on occasion and specifically mentioned aching legs and feet. Don Piccard recalls hearing this: "They say that when his feet bothered him, he would release pressure manually by the faceplate. They say that he did it repeatedly. And they argued against it." Yost heard Nick talk about the "tingling sensation" and a numbness in the face, and Janice recalls Nick mentioning that his face was extremely sensitive to the cold at altitude.

This all suggests a possible piece of the puzzle to Jack Bassick, executive vice-president and director of research and development for the David Clark Company. Bassick was a relative newcomer at David Clark while the company was cooperating with Project Strato-Jump, but he remembers those days well. He has since worked with the Air Force and with NASA on numerous projects requiring sophisticated pressure-suit systems. Bassick is not a physiologist, but he has some forty years of practical experience in the field of high-altitude physiology and protective equipment. And when he hears about a tingling sensation in the face, he recognizes the symptom.

"That's classic paresthesia," he says, "which is characteristic of decompression sickness. Absolutely classic." According to literature provided by the FAA's Office of Aviation Medicine, Civil Aeromedical Institute in Oklahoma City where Nick received pressure-suit training, symptoms of DCS may include "itching, usually around the ears and face" and the "sensation of tiny insects crawling over the skin." This phenomenon, sometimes called "divers' itches," is caused by tiny nitrogen bubbles trapped beneath the skin. But why would depressurizing the suit alleviate symptoms of DCS? How could discomfort caused by an external *decrease* in air pressure be alleviated—even temporarily—by *reducing* pressure inside the suit? Bassick speculates: "Dumping air out and having it refill with 100 percent oxygen—it sounds like this sort of refreshed him and made him feel okay. It might numb you a bit and you could become hypoxic very quickly. And if you're hypoxic you often have a euphoric feeling. You feel better. But you're just fooling yourself."

Raven Industries engineer Russ Pohl echoes Bassick's thoughts:

> Nick had the pressure suit on, and in those days the helmet had a dump valve right here [left side of the helmet]. If you got the bends a little bit you hit that dump valve and the pressure would drop and bring you right back up. . . . We postulated that he probably had the bends, hit the dump valve, and had these stiff gloves on. . . . And when that faceplate dislodged a little bit, he hit it with his glove and it didn't seal. That's where all that rushing of air came from. I mean, if it had punctured [somewhere lower on the suit], you wouldn't have heard it. But they heard it. It was very pronounced.

Suggestions that Nick may have been suffering from DCS during the flight of Strato-Jump III, and that these symptoms may have caused him to deflate the helmet's visor seal, bring us to the matter of the project's preflight denitrogenization procedures.

The proper interval for prebreathing 100 percent oxygen prior to a flight like Strato-Jump's is somewhat in question. Dr. John Paul Stapp's original denitrogenization experiments settled on thirty-minute prebreathing sessions. American high-altitude aircraft pilots typically prebreathe for one hour and fifteen minutes prior to takeoff. The Air Force and Navy manned balloon programs preferred even longer durations. The Excelsior project—the nearest operational equivalent to Strato-Jump—used two-hour prebreathing sessions. According to the reasoning of Joe Kittinger and his Excelsior team, it doesn't take long to flush the nitrogen from the bloodstream, but purging it from the body tissues requires longer intervals. An article by Dr. Bruno Balke of the Air Force's School of Aviation Medicine provides support for this line of thinking: "Breathing 100-per-cent oxygen for several hours before leaving for a high-altitude flight lowers the partial pressure of nitrogen in the lungs, blood, and body tissues *in that sequence*" (italics added). In Kittinger's opinion, for a stratospheric balloonist, anything less than an hour of prebreathing would be sheer folly.

NASA has on occasion required astronauts preparing for extravehicular activity to prebreathe for a full twelve hours to compensate for the differences between the spacecraft atmosphere and the EVA pressure-suit atmosphere, yet Soviet cosmonauts have typically been given only thirty minutes for the same exercise. While this extreme variation may be partially explained by differing suit designs and spacecraft-pressurization levels, it is clear that a consensus on proper prebreathing duration has never solidified.

It is difficult to reconstruct with any precision Nick's prebreathing activities on the Strato-Jump flights. According to most reports, he engaged in prebreathing of 100 percent oxygen in the suit-up area both during and after the suit-donning procedure. And while the equipment lists consistently include "walkaround" oxygen bottles for the suit-up area, because the preflight checklists don't specifically mention prebreathing prior to Nick entering the gondola, there can be no certainty about when Nick began to prebreathe or about how rigorously he continued it. What the checklists do routinely specify is a twenty-minute prebreathing session in the gondola immediately prior to launch. (Interestingly, the original Strato-Jump I flight plan assembled by Litton engineers and signed by both Tim O'Malley and Karl Stefan nearly two weeks before the Minnesota launch included a countdown script with the following notation at L - 1.5 hours [launch minus one and a half hours]: "Nick starts pre-breathing from auxiliary breathing supply while inside his gondola." Yet by the time Nick updated and published his own final pilot checklist a few days later, this item had been deleted. In Nick's checklist, the first mention of prebreathing occurs at L - 20 minutes.)

Prebreathing must be continuous, or very nearly continuous, to be effective. Because Earth's atmosphere is primarily nitrogen, regardless of the length of time spent breathing pure oxygen, the blood and tissues are renitrogenized in short order as soon as an individual resumes breathing air. On Project Excelsior, to guard against even trace amounts of nitrogen-rich air entering the lungs once prebreathing had commenced, Kittinger took extraordinary precautions while switching oxygen masks or donning his helmet. "All during this period," Kittinger later wrote about his preflight ritual, "I wore the oxygen mask; before donning the neck ring, Dan [Excelsior's Life Support Systems Specialist, Sgt. Robert Daniels] placed a clamp about my nose and I held my breath until the mask could be replaced. I didn't dare take one natural breath or I would jeopardize the entire denitrogenization process."

Jack Bassick notes that renitrogenization occurs on an extremely steep gradient. "You lose it [denitrogenization] very quickly," he says. "It only takes you a few breaths to put you back to square one. If you're out walking around and opening up [the helmet visor], opening up, opening up . . . you're back to square one."

Tom Smith, on site for the launch of Strato-Jump III, echoes this thought: "Usually pilots prebreathe for a flight like that for a considerable amount of time. It's completely against any kind of protocol to open up your visor. Once you start prebreathing, you have to stay prebreathing." Smith was well versed in the procedure because, as a field representative for David Clark on various Air Force programs, he was highly motivated not to become the cause of delaying or interrupting the procedure once it had begun: "You had to be there on time. The whole flight could be delayed if the prebreathing didn't start on time or was interrupted for any reason. Sometimes we would be involved in chamber testing with the Air Force pilots, so we would have to prebreathe ourselves. So we knew very well what the procedure was."

Given that prebreathing must be continuous or nearly continuous to be effective, it's puzzling why a twenty-minute session would even be specified prior to launch. One explanation may be that the twenty minutes of prebreathing in the gondola specified in the checklists referred to the minimum interval of prebreathing *after* switching to the onboard oxygen supply.

However, if there was any significant period between the time that Nick left the suit-up area and the time he began prebreathing from the gondola's oxygen supply during which he was breathing air, then any denitrogenization that had been achieved in the suit-up area would have been negated. In other words, if after leaving the suit-up area Nick did not continue to breathe 100 percent oxygen from the walkaround bottles 100 percent of the time, effective denitrogenization would be reduced to the mere twenty-minute session in the gondola immediately prior to launch.

When Dick Sears debriefed Dr. Thompson and the chase pilots, he inquired about the prebreathing process and included the following comment in his Strato-Jump III report: "He did pre-breathe for about 10 minutes but then he opened his visor for 2 or 3 minutes before lift off."

What is clear is that if the in-gondola prebreathing sessions (whether the duration was ten or twenty minutes) were all the effective denitrogenization Nick had prior to launch, then the process was too brief by almost any standard and he was physiologically unprepared to make the ascent. According to contemporary FAA guidelines, the prebreathing of 100 percent oxygen "for 30 minutes prior to initiating ascent to altitude reduces the risk of altitude DCS for short exposures (10–30 minutes only) to altitudes between 18,000 and 43,000 feet." Clearly, a balloon flight of three hours into the region above 100,000 feet warranted a longer interval.

The first Strato-Jump flight never got high enough for the suit to inflate and never experienced sufficiently reduced air pressure for DCS or denitrogenization to become an issue. And while Nick's reports of aching limbs and a tingling face indicate that he may have been suffering from some DCS symptoms on Strato-Jump II, the problem apparently didn't compromise that mission. Nor did Nick complain of lingering DCS symptoms after landing, which one would have expected had the in-flight symptoms been acute. So, was there anything different about the preparations for the final flight that suggest Nick might have been exposed to an increased risk of DCS?

According to Ed Yost something was indeed very different for Strato-Jump III. Yost says that when he showed up on the flight line in the predawn hours of May 1, he was shocked to see Nick, dressed already in his pressure suit but with his helmet visor open, sitting on the tailgate of the stationwagon that was used to transport him from the suit-up area to the gondola. Yost, who had no responsibility for the life-support systems but who was familiar with the drill from the previous flight, challenged Nick and asked why he wasn't back in the administration building breathing oxygen. That's when Nick replied that some members of his crew had failed to arrive at the appointed time.

"They weren't there," according to Yost. "That's what I was told, and I sure didn't see any of 'em. They certainly weren't there on the flight line when they were supposed to be there." Yost recalls Nick saying that he and Janice had worked together to complete the suit donning, although Janice does not remember this. Richard Sears's postflight report states, "I assisted Marv in the suit up of Nick. Everything went smoothly. Bill and Marv helped Nick get into his parachute harness and attach his emergency pack." Interestingly, Sears's report goes on to say that Nick left the administration building and went straight to the gondola with the balloon ready to go and the launch vehicle manned and waiting. This suggests that Nick had returned to the administration building and completed the suit-up/chute-up procedure following his conversation with Yost.

Although Yost wasn't completely familiar with the protocols for manned high-altitude flights, he knew enough to be concerned. But when he confronted Nick, Nick had been quick to reassure him.

"Even though he didn't have his prebreathing of pure oxygen," Yost recalls, "while he was sitting there on the stationwagon his spirits were really good. And he said, 'I've done this before.' I don't know what that meant. Probably that he'd made some high-altitude flights without it. Maybe he'd gone up before without prebreathing much oxygen. Maybe that's what he was referring to. But his spirits were good. He said, 'Don't worry, we're okay.'"

Earl Clifford remembers being surprised at Nick's approach to prebreathing that morning. Clifford knew high-altitude flight procedures, and Nick's lackadaisical attitude worried him. The prelaunch footage of Barry Mahon's Strato-Jump documentary *The Angry Sky* corroborates Clifford's and Yost's memories. Nick can clearly be seen approaching the gondola with the helmet visor open. This is essentially a smoking gun, proof that regardless of what did or didn't occur in the suit-up area that morning Nick took his seat in the gondola for the final time with only marginal protection at best against nitrogen expansion during ascent. And while it's clear that Nick did prebreathe in the gondola prior to launch, the interval has been reported as no longer than twenty minutes, and as short as ten. In spite of the high spirits on the ground and aboard *JADODIDE* early that morning, Strato-Jump III was a disaster in waiting.

One final question surrounds the mystery of what happened at 57,000 feet: Where did Nick get the idea that he could get away with even momentarily opening the helmet's pressure visor at altitude? Is there any circumstance in which such an action might make sense? According to Jack Bassick:

> He'd reportedly done it on previous flights. He said, "If I open my visor and get a couple of breaths of fresh air, I feel better." And a little hypoxia . . . makes you *feel* good! But he apparently did it at a lower altitude before where he didn't have a lot of suit pressure. And you can do that. The suit

> starts to pressurize once you get above 35,000 feet. So, below 35,000: no problem. That is, no problem from a suit-integrity standpoint. Of course, you lose any denitrogenization you've done up to that point. But you would never open your visor above 35,000 feet. Never. Ever!

An explanation for Nick's behavior may conceivably be traced back to his initial pressure-suit indoctrination at Tyndall Air Force Base a year earlier. On those decompression chamber runs, according to Air Force records, Nick had been instructed to briefly open the visor twice (once at 15,000 feet and again at 30,000 feet) in order to experience the symptoms of hypoxia for the purpose of being able to identify them should they occur on an actual high-altitude flight.

Bill Jolly insists to this day that Nick was only following instructions by opening the visor. "He was *taught* to do that," Jolly says emphatically. "When he went through his training down at the Air Force base, they taught him to do that. And even though he made the mistake of doing what he did, he was taught to do it."

Opening the visor at relatively low altitudes to expose yourself to hypoxia is part of the standard physiological training for pilots, and there should have been no confusion about the fact that the visor should never be opened once the suit is pressurized. If Nick or someone else told Jolly that he had been trained to open the visor above 35,000 feet, as Jolly says, then it's clear that a tragic lack of communication had occurred at some point. However, the tongue-in-cheek exchange between Nick and Marvin McCall on the second flight would suggest that Nick was in fact quite aware of the dangers of opening the visor in the stratosphere. During the ascent on that flight, Nick asked McCall how to get the perspiration off the interior of the visor and was told, clearly jokingly, to open the visor. When Nick answered that he would try to do just that, McCall countered quickly and soberly with the unequivocal command: "Negative."

And yet, another incident—this one during the descent of Strato-Jump II—indicates the possibility of some ambiguity in Nick's mind on this score. In his oral debrief following the flight, Nick mentioned something he had not shared with his ground crew at the time: "I couldn't open the facemask because it was frozen, which I knew it would be. [This attempt to open the visor likely occurred somewhere between 35,000 and 40,000 feet.] Once the suit depressurized, I did open the neck seal and raised the helmet up over my head. I was able to get some fresh air. There wasn't sufficient oxygen, and when I'd get a little woozy I'd have to pull my head back in." This suggests that Nick may have managed to open the faceseal by loosening the helmet take-up adjustment and hold-down strap which allow the helmet to rise up and cause a constant flow of oxygen across his face.

Jack Bassick again: "If he was properly trained—and we know what he did for training and qualification—he should have known better than that. I guess that's the surprising thing. Somehow it seems as if he didn't get it." Certainly the operation and service manual for the A/P22S-2 pressure suit is unequivocal on this point. The following entry appears on the manual's final page, italics included: "CAUTION—Never open visor while suit is pressurized *regardless* of altitude."

A Credible Scenario

More than thirty-five years after the fact, the surviving physical record of what occurred at 57,000 feet aboard *JADODIDE* is woefully incomplete. This has obviously added to the historical confusion surrounding the flight. Yet some of the evidence that many have assumed could settle the matter once and for all probably never existed in the first place.

Barry Mahon had two movie cameras on board: one positioned beneath the gondola, pointed toward the ground, to capture Nick as he jumped, and another inside next to Bob Kelley's still camera at the top of the gondola. Both of Mahon's cameras were battery-operated World War II–era models and both were covered in heat tape. Because everything that happens in skydiving happens fast, these cameras—like most used to capture free-fall action—were overcranked to forty-eight frames a second to produce slow-motion footage. For this reason, the cameras' one-hundred-foot loads would yield a little less than three total minutes of finished film, requiring a judicious use of the manual trigger that controlled their operation. Nick did run off a few feet of film during the launch sequence on Strato-Jump III, but due to a desire to save most of the film for float altitude and for the jump itself, he probably didn't trigger either camera again that day.

Reports have emerged over the years in which unidentified individuals claim to have seen film of Nick slumping forward following the decompression at 57,000 feet, but given that Nick would have had to trigger the interior camera immediately before or during the emergency, such reports are highly suspect. Likewise, the still cameras on board required manual triggering, and it is difficult to imagine Nick snapping photos during his struggle. Finally, Phil Chiocchio, who was part of Mahon's crew and who helped edit the raw flight footage for inclusion in what became their tribute to Nick, *The Angry Sky*, recalls seeing no in-flight film (beyond the few seconds shot during liftoff) of the May Day attempt—and none whatsoever of the incident at 57,000 feet. It's doubtful one would forget that, even thirty-five years later.

On the other hand, good quality audiotapes of the ground/air communication were made of all three Strato-Jump flights. Ben Raiche, a Raven employee who supervised the company's test laboratory, was responsible for operating the reel-to-reel tape machine for both of the Sioux Falls flights and for making backup copies following the flights.

When a tragedy like Nick Piantanida's occurs, even in the relatively nonlitigious days of the mid-1960s, the fear of lawsuits descends on all parties involved, on companies as well as individuals. Because of the mystery surrounding the incident, and in spite of the fact that Nick's files included a stack of signed liability wavers—just about everyone and every entity who agreed to cooperate with him required a signed release—most everyone was to one degree or another worried about getting sued. This paranoid climate was exacerbated by reports of threats from unidentified members of the Piantanida family in the days and weeks following the accident, although neither Vern nor Janice can recall any such threats. While cool heads ultimately prevailed and no legal actions were ever filed, none of the participants felt completely safe, and no one was especially anxious for the voice tape of the final flight to become part of the public domain.

Ben Raiche recalls making two copies of the original tape. One copy went to Ed Yost and another was given to Dick Sears who took it back to the David Clark Company plant in Massachusetts. The original was kept in a file cabinet in the Raven Industries building. While the original and both copies were thought to have been destroyed years ago, a third copy was discovered during the research for this book. Using that tape, we can construct an accurate chronology of the final moments of the flight of Strato-Jump III.

JADODIDE was launched at 5:45 A.M. As he began his instrument check at 50,000 feet, Nick gave the time as 6:34. The instrument check was followed by one of Nick's lengthier transmissions during which he described the rotation of the gondola and his view of the landscape. For approximately the next minute, Nick and Darrel Rupp discussed the gondola's position and the fact that it was moving eastward faster than anticipated.

6:40:50 *Darrel Rupp:* "Nick, Dick Keuser said that you've slowed down and are moving very little now."

6:40:55 *Nick:* "Roger." [Nick actually says something like: "Ohh-Roger." It was a mannerism that was new for the third flight. His voice sounds quite comfortable here, relaxed and unconcerned.]

6:41:13 *DR:* "Okay, Nick. On your next minute, would you drop another one minute of ballast?"

6:41:19 *NP:* "Roger." [Since we don't know precisely when the next minute began on Nick's onboard clock, we don't know whether he ever hit the ballast switch that final time. If he hit the ballast-open switch, we can be fairly certain that he never had an opportunity to close it.]

6:41:27 *Al Tomnitz:* "Ed, I put him right over Worthington. About three miles west of Worthington at this time."

6:41:34 *DR:* "Roger. Thank you, Al."

6:41:40 *AT:* " . . . goin' on the ground there." [During the eighteen-second interval immediately following Tomnitz's incomplete comment, Nick inhales/exhales four times. His breathing sounds harsher here, more labored than at any point during the flight.]

6:41:57 [A very brief sound that is best described as a light rumble, as if Nick had pursed his lips and blown into the microphone—unlikely since he was just beginning to inhale.]

6:41:58 [This is the instant when the "whoosh" is heard. It has an audible duration of three seconds and tapers off in volume and intensity. It sounds like "shhhhh." It is very distinct, but not especially loud—not as loud as the sound of Nick's breathing has been to this point. It is hard to imagine the sound as anything other than air escaping from a bladder.]

6:41:59 *NP:* "Close it!" [This is the vocable that no one could understand. Janice thought that it might have been Nick sneezing.

Somebody at Raven believed at first that Nick had said, "Visor!" But after repeated listenings on sophisticated audio equipment, the best guess is "Close it!" or "Close 'er!" Nick says it before the hissing sound has stopped. His voice is loud, urgent, and perhaps slightly slurred.

The sonic atmosphere is markedly very different just at that moment, very much as if Nick is speaking into a strong wind. In the few seconds following Nick's words here, the familiar sound of his breathing seems at first to have stopped. On closer inspection, he can be heard to take—or attempt to take—two or three very short, very faint breaths.

Was Nick addressing the ground crew? There was nothing they could do for him at that moment, nothing for them to "close." The comment may have been directed at himself if he was, as seems plausible, struggling to reseal the helmet visor.]

6:42:08 *DR:* "What was that, Nick?"

6:42:10 *NP:* "E*mer*gen . . . !" [Nick's voice is loud here, and distinctly urgent, but it is not a scream and is nowhere close to hysterical. His tone is a bit choked, almost strangled, as if he is working hard to force the word out. The word cuts off abruptly before the final syllable is heard—followed by dead air. The elapsed time from launch to Nick's final transmission is almost exactly fifty-seven minutes.]

Relying on that voice tape and on the remainder of the information we have—some of it conflicting, much of it speculative—what are we left with? Can we construct a credible scenario of what happened on May Day morning, 1966?

First, what we know:

Nick's body had not been sufficiently denitrogenized for a stratospheric flight of more than two hours. Nick had opened his visor for some significant period of time after exiting the suit-up area at Joe Foss Field and before taking his seat in the gondola. (This is verified by film footage from *The Angry Sky* and by generally reliable witnesses.) Therefore, he had a maximum of twenty minutes prebreathing prior to launch, and quite possibly less. We have reason to believe that Nick had suffered from some degree of DCS on Strato-Jump II because of his reports of facial-skin tingling, and the denitrogenization activities prior to Strato-Jump III afforded him, if anything, an even shorter interval of prebreathing.

Next, what we can be fairly certain of:

Nick opened the visor in flight, probably deliberately. Nick's visor was open when Dr. Thompson and the two Raven pilots arrived at the landing site. (This is based on Thompson's report to Dick Sears the day following the accident, and there is no compelling reason to doubt its veracity.) Given the near impossibility that the visor could have opened by itself because of the powerful spring-and-ratchet control and pneumatic seal that locks it in the closed position, we can be fairly certain that Nick opened the visor. (It seems likely that the "whoosh" at 57,600 feet was a result of the suit's oxygen rushing past the open microphone at Nick's lips and out through a breach at the visor seal.) We also have Ed Yost's statement that Nick had mentioned deliberately depressurizing and repressurizing the system by depressing the visor bearing, along with Bill Jolly's assertion—correct or not—that Nick had been trained to do precisely that.

Neither the pressure suit nor the helmet malfunctioned. The exhaustive inspections and tests performed by the David Clark Company, initially in the presence of and assisted by experts not associated with the company, were unable to identify any problems with the suit system. Again, the helmet would be the most likely culprit given the rush of air heard on the ground. Yet the David Clark Company has no record of any catastrophic failures with that model of helmet throughout its history. The surviving photographs that Dick Sears took of the David Clark equipment in the Raven labs after the flight show an apparently undamaged helmet with an intact pressure visor.

Nick began an attempt to exit the gondola shortly after the rush of air was heard. Because his vocalization (his attempting to transmit the word "emergency") was terminated so abruptly, and because his communication line was disconnected when Dr. Thompson and the pilots arrived at the landing site, we can be fairly certain that Nick purposely pulled the communication connector. The main line from the gondola's oxygen supply to the suit was also disconnected when Dr. Thompson arrived, suggesting that Nick pulled the quick-disconnect. The bailout oxygen cylinders were empty when Dr. Thompson arrived, suggesting that the bailout supply had been vented to the air through the open visor and had been exhausted during the long descent. So, in spite of the fact that Nick didn't follow all of the proper emergency procedures—he didn't, for example, jettison the balloon himself—we can be fairly certain that he was preparing to leave the gondola in flight when he was overcome by hypoxia.

What we can assume given the preponderance of evidence:

The oxygen supply was not sufficiently contaminated to have caused a problem and the oxygen delivery hardware did not malfunction. Not only were multiple checks of the oxygen supply and delivery hardware performed in the days and hours prior to launch, but the oxygen system behaved flawlessly until the rush of air was heard. In addition, the oxygen system was tested, along with the suit and helmet, in both pre- and postflight tests at Raven Industries, and no anomalies or contaminants were found. Had there been a problem with the oxygen supply or delivery system, it would have been very much in the interest of the David Clark Company reps to have discovered that—and they did not. It is instructive to note that even had contamination of the main oxygen supply caused an interruption of oxygen flow, an alternate supply—the bailout bottles in Nick's backpack—were available and easily activated.

Finally, something we don't know:

There were at least six individuals present at the launch of Strato-Jump III who understood prebreathing and knew the proper procedures. Why didn't someone intervene? The protocols of the prebreathing process were not necessarily common knowledge even to the experienced balloon operations personnel, most of whom had never before been associated with a manned high-altitude launch. Karl Stefan's question to Nick, well into the flight of Strato-Jump I, asking whether his pressure visor was closed, is a case in point. But on May Day morning at Joe Foss Field, the knowledge was there. One of Raven's employees had been a military jet pilot who had prebreathed himself; the representatives from David Clark and from Firewel knew the process intimately; Dr. Thompson had aeromedical training and certainly should have known; and Marvin McCall was in the business of training high-altitude pilots.

Nick reportedly walked around with his visor open or his helmet off altogether *after* completing his prebreathing in the suit-up area. Janice was there for all the launches, and she remembers it this way: "He would be away from me, and it was to prebreathe oxygen. And then when he got into his suit, he was without the oxygen. And that's when I stood next to him and they didn't let me touch him. And then I kissed him goodbye, *and then he'd put his helmet on* [italics added]."

But according to Tom Smith of the David Clark Company, "There are requirements for prebreathing. Once you start prebreathing, you have to stay prebreathing." And Smith can't imagine seeing a high-altitude pilot walking around breathing air thirty minutes before taking off. "Anyone—Dick or I or any of the other FSRs [field service representatives]—if we had observed anything like that, we would have made an issue of it." Yet Dick Sears was told after the flight that Nick had in fact opened the visor for two to three minutes after taking his seat in the gondola.

So why didn't anybody say anything? The only answer may be that the crowded, somewhat chaotic conditions around the launch site and the gondola that morning made Nick's activities difficult to observe for some of the key participants.

In the end, a reasonable scenario, in fact the only believable scenario that fits the evidence, is that Nick opened the pressure-sealed helmet visor, that he deliberately took the action that resulted in both a loss of oxygen and rapid decompression. And while Project Strato-Jump's inadequate rigor in the performance of the preflight denitrogenization procedures suggests that symptoms of decompression sickness may have been the cause of Nick's puzzling and ultimately fatal action, regardless of *why* he did it, we must conclude that his own actions—and not equipment malfunction—were to blame.

Legacy

In spite of the fact that Nick was deadly serious in his approach to Project Strato-Jump and in spite of the fact that he understood—in the main—the dangers inherent in stratospheric flight, nothing in the preceding scenario is incompatible with Nick's previous behavior or what we know about his character. Raven's Russ Pohl thinks it was simple carelessness. "Nick was always a cocky guy," he says.

Roger Vaughan has a similar take on the flaw he saw in Nick Piantanida: "The flaw was this slightly over-the-top kind of attitude, where everybody was trying to hold him back." Vaughan goes on to liken Piantanida to some of the Air Force legends who'd preceded him: "But that flaw . . . all of them had it to some degree. [Dr. John Paul] Stapp had that flaw. Kittinger had that flaw. Any of these guys could have stubbed their toe on their own selves. And that's what Nick, I think, did."

Yet as Vaughan looks back on the Strato-Jump days (without a *Life* editor rearranging his conclusions), he views Nick as the last of a dying breed:

> He was always a little bit over-eager, a little bit on the reckless side. Full head of steam, pedal to the metal. That was Nick. If you went with Nick to have a beer, you'd have six beers. Life to the fullest. That's the way he did everything, that's the kind of life he lived. I mean, if you're going to go climb Angel Falls you can get down there and take a look at the thing and think of fifty reasons why you're not really ready to do it. But you wouldn't be Nick if that were the case. And I think today that's really hard for people to understand, because we're in such a safety-mad universe.

Recent efforts to decode the human genome have given us new insights into the composition of personality. Some of this work may shed light on extreme characters like Nick Piantanida. On chromosome 11 of the human genome is a gene known as D4DR which constitutes a recipe for a protein known as a dopamine receptor. Dopamine is believed to be one of the brain's principal "motivational" chemicals. Experimental evidence suggests that individuals with "long" copies of D4DR have a lower responsiveness to dopamine and therefore require more of the chemical than those with "short" copies. Long D4DR types tend to be novelty-seeking, adventurous, and easily bored; short D4DR types are more shy, inhibited, and disinclined to action, change, or risk. On the basis of such theories, we would expect Nick to have had some of the longest D4DR around.

But because—as researchers believe—environment has such a profound effect on biology (and especially on chemical reactions in the brain), it is possible that the highly competitive social life Nick experienced growing up on the streets of Union City was at least as influential on his personality as brain chemistry. While it might be overly simplistic to describe Nick as a dopamine-riddled extrovert whose outgoing character was exacerbated by an aggressive social world, it is probably fair to say that a rare combination of brain chemistry and environment resulted in the extraordinary product that was Nick Piantanida.

But Nick was never a simple character. He could move gracefully through opposing worlds. He was the guy with the cobra in the Ichi-Ban, but he was also the guy with the briefcase and his own office space on Capitol Hill in Washington, D.C. With only slight alterations, or perhaps mere shadings, of accent and diction and delivery, he could navigate quite naturally between the beer and brawls of Jim Lee's barbecue joint in Sioux Falls and a general's private jet. "He had a great sense of humor," according to Ed Yost. "And he was very serious, too. He could be both."

In the estimation of most who spent time with him, he was a great guy to know and to be around. Yet few knew him at all well, a paradox in large part attributable to his difficulty with compromise. With neither apology nor egregious defiance, Nick was willing to apply himself to only one thing: his own ambitions and dreams. He was interested in only one destination: a place you couldn't follow.

Whether one believes that the mid-1960s witnessed an unraveling of civil society or the birth of a new national culture, America had experienced profound changes in the Strato-Jump years. At the time Nick made his first parachute jump at Lakewood, John F. Kennedy was still in the White House, Vietnam seemed little more than a foreign policy annoyance, and no one in the United States had ever heard of the Beatles. In one sense, Nick's odyssey had taken root—spiritually, if not literally—in the faded optimism of the 1950s. By the spring of 1966, not only fate, but the times, had caught up to Nick. On the morning of the day on which he set the unofficial manned-balloon altitude record, the *New York Times* carried on its front page two stories about seemingly unrelated events—one, at that time, a relatively novel act of defiance, the other a continuation of a long struggle—that hinted at things to come. In New York thirty-two demonstrators had halted traffic in Times Square during a sitdown protest against the bombing of civilians in North Vietnam, and in Greenville, Mississippi, 110 civil rights workers and poor blacks had occupied a deactivated Air Force base and were forcibly evacuated. On the day of Nick's final flight, the *Times* carried a photograph of a slumped, disconsolate President Johnson with the caption, "Problems, Problems." The changes—in America, in Western culture—would erase Strato-Jump from memory, as if a new collectivist radar was incapable of even discerning the solo act. Nick Piantanida had the misfortune to be cast as a pioneer just as the pioneering days, the cowboy days, of upper-altitude and space exploration were drawing to a close.

Nick had gotten his one golden chance on Strato-Jump II and had been defeated by the seemingly insignificant: a simple disconnect fitting. "If only I'd had a damn dollar and twenty-five-cent wrench!" Or perhaps he was merely defeated by the illusion that *anything* at 123,500 feet could be insignificant. If Nick's dream had only been a shade more prudent, less grand—if he could have whittled it back on the fly, if he'd been content to settle for the unofficial lighter-than-air world altitude record and been able to share Jacques Istel's satisfaction in that achievement either as it was happening or in retrospect—he might have found a measure of pride in accomplishment, his measure of success. Nick's currents ran deep, and in some ways he'd been successful in walling himself off from potential distraction and retaining an unyielding focus on the business of Strato-Jump, at least through the second attempt. But the deeper waters eventually began to exert their force in the late winter and early spring of 1966. The concerns of family (the choice of *JADODIDE* as the name of the gondola for the final flight), thoughts of his own mortality (his conversation in Jim Lagomarsino's ravioli shop), and his own impatience for glory intruded on the feasibility of the dream. Far from diminishing Nick, this inner conflict humanized and in some ways strengthened him. And yet Strato-Jump III found only doom, a doom that was tied to all the nagging failures that had come before, the failures that seemed, in the end, to eclipse all the achievements of Nick Piantanida.

Project Strato-Jump has had its imitators and will eventually have its successors. The money, the will, the know-how, and the luck will one day come together. But it can never happen in quite the same context again. For Nick it was never about mere thrills or kicks, which speaks to the difference between Nick (at least the Nick of the Strato-Jump days) and most of the "adrenaline junkie" types of today. Certainly we will never be able to think about such projects the same way. A November 2001 *Scientific American* article on twenty-first-century aeronauts who desire to break Joe Kittinger's free-fall record and to break Nick's absolute balloon altitude record is openly cynical about the value of such attempts: "Undoubtedly, their efforts will generate data about stresses the human body can—or cannot—endure. But then again, so does MTV's *Jackass*." The latter-day projects have brought some of this on themselves.

From one perspective, today's would-be stratospheric explorers whose projects will require not tens of thousands but several millions of dollars in funding have simply taken the ad hoc public relations efforts of Nick and Jacques Istel to a higher, or at least slicker, level. But from another, the clever Web sites and Hollywood-inspired collateral of today's projects seem to trivialize high-altitude exploration even as they attempt to invest it with meaning, transforming their efforts into quasi-inspirational marketing campaigns. The obligatory scientific justifications are still present, but a certain New Age sensibility that would have baffled the Strato-Jump team often dominates. One such project's advertisements talk about "sustenance" and "vision," about "entrusting a soul life into the hands of a Chosen Team." Another casts a free-fall record attempt as an antidote to "deviant places of refuge" like religious cults, hard drugs, and delinquency. In yet another stratospheric skydiving project's list of goals, the word "safety" appears in the first sentence of each of the four objectives.

Not only are the Nicks gone, their time is gone as well.

According to the original business plan drawn up in Jacques Istel's midtown Manhattan office in 1965, the purpose of SPACE, Inc. had been to sponsor a series of experimental high-altitude parachute jumps. These jumps would be designed to "seek out, test and improve the methods of high altitude bail-out survival." Project Strato-Jump—envisioned as just the first of several such projects—had four quite specific objectives:

1. To establish that a trained parachutist can free fall from altitudes in excess of 100,000 feet without the use of stabilizing devices.
2. To investigate the effects of transonic speeds on the human body in free fall.
3. To gain for the United States the world's free fall altitude parachuting record which is officially held by Russia.
4. To surpass the manned balloon altitude record.

In the end, of course, the project satisfied only the final objective: Piantanida and company surpassed the manned balloon altitude record of Malcolm Ross and Victor Prather of the U.S. Navy in *Strato-Lab V*. Ross and Prather reached 113,740 feet on their project's final flight in 1961. Strato-Jump II had topped out at 123,500 feet, by any measure a remarkable achievement. If in practice we accept Joe Kittinger's epic free fall as a world record—and mostly we do, in spite of FAI and NAA requirements—then we must accept Piantanida's altitude record as well. There is no disputing the fact that Kittinger *did* jump from the Excelsior gondola at 102,800 feet, or that Piantanida *did* take the Strato-Jump gondola to 123,500.

Nick is only the fifth person—and the lone civilian—ever to surpass 100,000 feet in a balloon. His attempt with Strato-Jump III would have made him the first balloonist ever to make it above 100,000 feet twice. His record (unofficial though it will always be) has remained essentially unchallenged for more than thirty-five years. It is one of the—if not *the*—most enduring altitude records in the more than two centuries of lighter-than-air flight. And even though Strato-Jump made no significant contribution to high-altitude research, Nick's dream has found its own peculiar niche in aeronautical history. The analysis of Raven's Jim Winker is perceptive: "It has to be looked upon as an attempt of the human spirit to do something remarkable. And I respect that. It takes a lot of guts to do something like that, and in the end that probably gives me my best feeling for Nick, that he had the guts to try it."

In June of 1966, barely a month after the flight of Strato-Jump III and with Nick still in a coma in Minneapolis, *Parachutist* magazine published a letter from a veteran jumper volunteering to carry on the attempt to break the Russian record and suggesting that the project be rechristened "PIANTANIDA 4." But by that time Hal Evans, an experienced parachutist and associate of Nick's who had served on Project Strato-Jump as an official observer for the Parachute Club of America, had already announced his own intention to complete Nick's mission with a jump from 120,000 feet. Evans's attempt, it was announced, would be made in honor of his fallen colleague and would employ much of the same equipment and personnel Strato-Jump had used. Evans intended to launch from Joe Foss Field on December 3.

But, as Roger Vaughan observed, there are always fifty reasons not to do it. And Hal Evans wasn't Nick Piantanida. The follow-up jump never materialized.

Strato-Jump was not only the last of the great civilian high-altitude balloon projects, it represents the last manned balloon journey into the upper stratosphere, period. Yet there is some evidence that Strato-Jump had something to do with its ultimate status, that Nick Piantanida's tragedy had a chilling effect on subsequent manned flights above the tropopause. There are those in the American ballooning and parachuting communities who continue to blame the project's failures for intimidating would-be partners and benefactors.

One 1990s American venture that had set out to follow in Piantanida's footsteps in an attempt to break the free-fall records of both Kittinger and Andreev cited Strato-Jump pejoratively in its project press release:

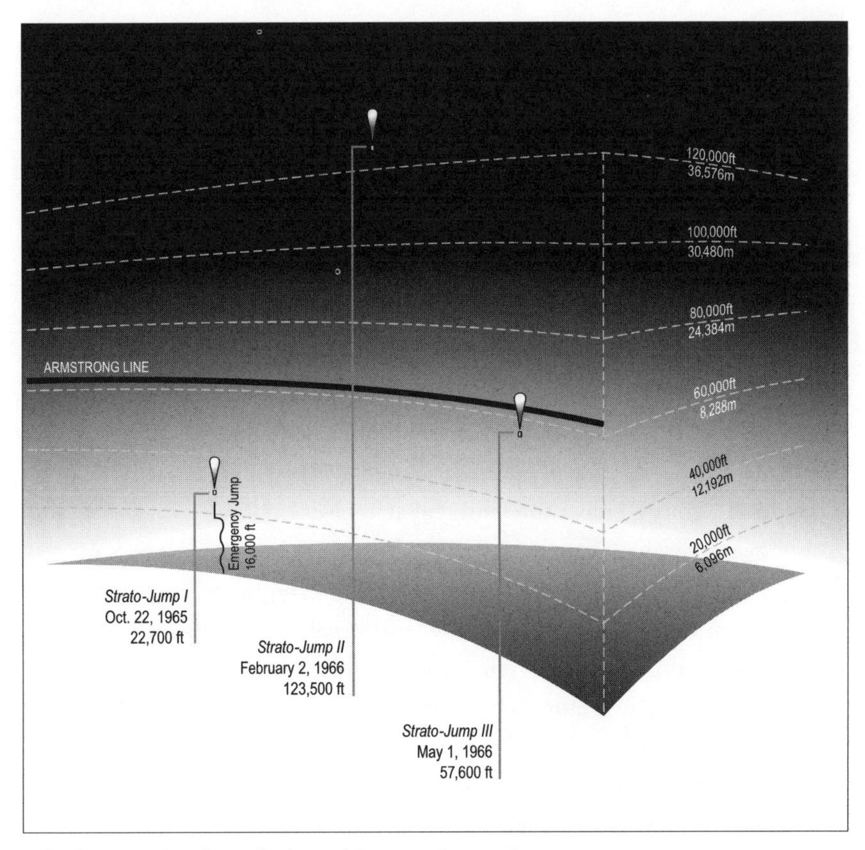

The legacy: the three flights of Project Strato-Jump.
(Courtesy Jonathan Townsend Lee)

> Examining the history of Project Stratos [*sic*] with its three flights, one comes to the conclusion that the lack of planning and preparation in all phases of that endeavor made its outcome almost inevitable. . . . The stigma left by Project Stratos in the United States aviation and aerospace community represents a considerable obstacle to any subsequent efforts to challenge these aviation records, and is in our opinion the reason these records have stood to this day.

This rhetoric seems unduly harsh. Certainly, contractors like the David Clark Company did become even more rigorous in evaluating potential participation in private aeronautical activities following Strato-Jump. Jack Bassick explains: "It certainly was a sobering experience. And it made you think long and hard before you ever got involved with anything like that again." The company eventually came up with the equivalent of a form letter as a response to the many civilian inquiries it receives each year. The letter stresses the dangers of low-temperature, low-pressure environments and explains that pressure suits are custom garments, all built to order, and hence quite expensive. It also points out that extensive qualification and testing are essential to any high-altitude manned operation, and that the assistance of government agencies such as the Air Force and NASA is crucial since only they have the extensive facilities and equipment required to properly train high-altitude pilots. But David Clark did not cease assisting qualified, well-prepared individuals in need of survival equipment. In fact, the company continues to cooperate with serious, record-seeking civilians who demonstrate their ability to meet the stringent standards imposed.

Following Strato-Jump, both the Air Force and the Navy, along with veterans of the military balloon programs, surely strengthened their resolve to refrain from encouraging would-be recordbreakers who would almost by definition be unprepared and unequipped to attempt—much less survive—such dangerous flights. However, the truth is that the successes of the American space program—and not Project Strato-Jump—killed manned stratospheric ballooning. NASA sent Alan Shepard into space a mere twenty-four hours after Ross and Prather's world-altitude-record flight for the Navy. When all was said and done, plastic gas bags at twenty-one miles had absolutely no chance against liquid-fuel rockets carrying human cargo aimed at the moon.

Piantanida and his project did plenty of planning and preparation. What they clearly did not do enough of was testing. The hardware and the pilot could both have benefited from more time in decompression chambers at extreme temperatures. More testing of the quick disconnect in a chamber at 70 degrees below zero, for example, might have turned up the problem that forestalled success on Strato-Jump II. One crucial difference between government-funded programs and civilian efforts is the aggregate brain power available. As Jim Winker points out, with civilian projects "you simply don't have enough people thinking about what can go wrong, and therefore things get overlooked." High-altitude manned ballooning is a brutal and costly business. Joe Kittinger has joked that the ideal way to prepare for a super-high-altitude balloon flight is to climb in the shower with your winter clothes on, turn on the cold water full blast, and spend a few hours stuffing $100 bills down the drain. But it's only half a joke.

Ed Yost, who has participated in hundreds of experimental balloon flights—both manned and unmanned—and who has been highly critical of many, judged Piantanida to be both serious and responsible: "I think he thought it out very well and I think he prepared himself very well." Yost assesses Project Strato-Jump this way: "The only problem they had was finances. They didn't really have the money. But everything was done as good as it could possibly have been done with the limited finances that were available."

Piantanida's death was hard on Yost, who knew from the start that Strato-Jump was a risky operation, underfunded and at times a little out of control. But like most of those who had gotten to know Nick, he judged the younger man to have a fighting chance. What's more, he thought Nick *deserved* that chance. The May Day accident got Yost where it hurt most. He lost a good friend, but also—perhaps—a piece of a dream that he had come to share. And it certainly did sour him on cooperating with future high-altitude projects.

"When you help somebody on something like this," Yost explains, "you work together and you get pretty close. He's like a brother to you. And then this kind of thing happens. . . . It's pretty tough. I dreamed about it."

Thirty-five years later, talk of Nick Piantanida still elicits raw emotion. "He was the kind of guy you'd like to have for a friend. And if you were a relative or brother or cousin or something, you'd be damn proud of him. A great person," Yost says wistfully. "A great person."

While almost everyone in Sioux Falls had liked Nick personally, not everyone concurred with Yost's assessment of Piantanida's preparation and discipline. John Kittelson was the project engineer at Raven and was responsible for overseeing and coordinating the engineering and manufacturing of the balloon for the Strato-Jump flights. Kittelson had been a fighter pilot and had flown the F-102 into the stratosphere. He understood high-altitude flight and prebreathing, and appreciated both the difficulties of survival in the stratosphere and the discipline and training necessary to avoid injury or worse. From this perspective, Piantanida and his mostly civilian crew appeared to be out of their league.

"He seemed confident from his previous parachute drops," according to Kittelson, "but I had the feeling that he was not professionally *there*. And it bothered me to see somebody smoke a few cigarettes and go jump off into space. I thought it was a pretty casual attitude." Kittelson also questioned whether the Strato-Jump crew was really qualified to advise Nick on super-high-altitude flight. Even the best technicians, he points out, are by definition almost totally focused on the proper operation of specific equipment. Who focused on the suitability and readiness of the pilot?

But Kittelson recalls his pride in the performance of the Raven operations team: "Good launches in both cases. But of course they'd been sending up very expensive scientific instruments for years. I think as a company we felt like we delivered what we were supposed to do. And we did it well. Among a number of us, an unofficial consensus was that he goofed up somewhere. Hit the wrong button."

Ed Yost and Jacques Istel contributed to the historical consideration of Piantanida's magnificent dream by shipping the Strato-Jump gondola to the Smithsonian Institution that summer. It remains in the Smithsonian's custody today.

Supporting Cast

A little more than ten years after the tragic end of Project Strato-Jump, in October of 1976, Ed Yost launched solo from Milbridge, Maine, in a self-designed, self-built gas balloon called the *Silver Fox*. He drifted across most of the Atlantic Ocean. After a flight of 108 hours, his boat-shaped gondola—which Yost preferred to a pressurized capsule that would have provided a more comfortable journey and quite possibly a better chance of success—splashed down 700 miles short of the Portuguese coast, setting duration and distance records for that class of balloon that still stand today. In 1978 a Yost balloon called the *Double Eagle II* piloted by Ben Abruzzo, Maxie Anderson, and Larry Newman completed the first successful crossing of the Atlantic, and in 1983 Joe Kittinger flew a Yost balloon called the *Rosie O'Grady* in the first successful solo crossing.

Yost holds six world records in helium balloons, twenty-one balloon-related patents, and has won numerous aeronautics honors and awards. During the Vietnam War he designed and built 87,000 rapid-deployment cargo parachutes for the United States military. In 1995 he became the first living inductee into the International Balloon and Airship Hall of Fame. Ed Yost currently runs his own companies, Universal Systems, Inc., and Skypower, and lives with his dog Shadow beside a trout stream in the high mountains of northern New Mexico.

Now busily retired from Raven Industries, Jim Winker remains one of the world's authorities on plastic balloons and parachutes and on the performance of lighter-than-air devices at extreme altitudes. He holds more than a dozen patents related to balloon systems design and engineering. A balloon pilot for more than forty years, Winker has flown in festivals and competitions all over the world. Now a noted lighter-than-air historian, Winker runs his own consulting business in Sioux Falls and conducts research for NASA.

In 1974 Jeanette Piccard became an Episcopal priest, one of the first women to be so ordained. She died in 1981. Don Piccard has held world altitude records in five different balloon categories, one of which remains on the books today. He has served as president of the Balloon Club of America (later reorganized as the Balloon Federation of America), and—along with Karl Stefan—founded the Plastic Balloon Club (later the Aerostat Society of America), and has written for numerous aviation-related publications. He founded Don Piccard Balloons, Inc. after leaving Raven Industries. His ambitions are as impressive as his accomplishments. Piccard's Project Pleiades aims to set an absolute manned-balloon altitude record using a cluster of the Mylar balloons that Piccard has always preferred. Piccard hopes to reach an altitude of 150,000 feet, more than four miles beyond Piantanida's mark. As an acquaintance observes, "Don is very interested in the Piccard tradition of being famous."

After losing his job at Litton Industries following the balloon failure on Strato-Jump I, Karl Stefan went to work for the National Center for Atmospheric Research, first in Boulder, Colorado, and subsequently in Palestine, Texas, and later worked for the International Latex Corporation, which became a pressure-suit contractor for NASA. He also worked for Tracy Barnes's balloon-manufacturing company, Balloon Works, where his office was a cell in a former women's prison on the campus of the University of North Carolina. Stefan has served as a consultant on a number of manned balloon flights and has remained active in the sport ballooning community.

He currently serves on the board of advisors for Don Piccard's Project Pleiades and lives with his wife, Lucy, in the mountains of Colorado.

Unable to get himself assigned to NASA's astronaut selection pool following Project Excelsior and Project Stargazer, Joe Kittinger volunteered for duty as a fighter pilot with the Aero Commandos in Vietnam where he would ultimately log over 1,000 hours in combat. In 1972 he was shot down over North Vietnam, captured, and imprisoned for eleven months in the notorious Hanoi Hilton prison camp. It might be easier to list the medals and honors Kittinger hasn't won than to list those he has, but a short account of those he has been awarded would include the Distinguished Flying Cross, the Bronze Star, Silver Star, Purple Heart, Legion of Merit, the Republic of Vietnam Cross of Gallantry, and induction into the Aviation Hall of Fame. He served as flight director for Ed Yost's transatlantic solo attempt in 1976. In 1983, in the first successful solo crossing of the Atlantic Ocean by balloon, Kittinger launched from the Maine coast and landed in a tree in the mountains near Savona, Italy.

Not long after Piantanida's final flight in 1966, Kittinger was contacted by the surgeon general of the U.S. Air Force, who expressed his gratitude that Kittinger's stubborn refusal to cooperate with Nick had kept the Air Force away from any official connection with the disaster on Strato-Jump III.

As a result of a program of self-administered medication prescribed for pain following surgeries on his pancreas, Jacques Istel remained a Demerol addict for the duration of Project Strato-Jump. While he was an energetic, vital, and creative contributor to the project from the beginning, he retains few lucid memories of that time. He was, for example, present at all three launches—but recalls none of them.

After selling Parachutes, Incorporated, to pursue other business opportunities, Istel moved to the desert in the extreme southeastern corner of California within sight of the Mexican border and founded the town of Felicity, named after his wife, Felicia, in 1986. He has subsequently had the location legally declared "The Official Center of the World." As the town's indomitable mayor-for-life, Istel has dedicated much of his energy to erecting a phalanx of granite walls commemorating significant individuals and historic events. One wall, for example, bears the names of all the Marines and Navy corpsmen who died in the Korean War. Another is dedicated to the history of French aviation. Istel envisions hundreds of such walls and already has architectural drawings showing their placement. One future wall will record names for the Hall of Fame of Parachuting. That wall must, of course, include the name of Jacques Istel himself.

Phil Chiocchio and Dan Quinn were rolling through North Carolina in one of Barry Mahon's film equipment vans, on their way to Miami to shoot a picture, when news came over the radio that Nick had died. Phil, who would be starting college that fall, recalls being overwhelmed by the sense that an era had passed. "It was an eerie moment. It was a sad time that has affected my life since."

Barry Mahon's busiest year as a director and producer was 1965. His cinematic output slowed somewhat during his involvement with Project Strato-Jump but picked up again after Piantanida's death. In the year following Strato-Jump, Mahon's credits included *The Sex Killer*, *Good Times with a Bad Girl*, and *Fanny Hill Meets Dr. Erotico*. Mahon would eventually become something of a cult legend as the King of the B-Movies. He died of heart failure in Las Vegas in 1999 at the age of seventy-eight.

Sen. Harrison Williams resigned in disgrace in March of 1982 just before the U.S. Senate was to vote on a resolution to expel him for his role in the Justice Department's ABSCAM sting operation. Williams offered to use the power of his office on behalf of a titanium venture controlled by what he was led to believe was an Arab sheik. Had he not resigned he risked becoming the first senator to be expelled by his colleagues since the Civil War. Williams was found guilty of bribery and conspiracy in 1981 and served two years in prison. He died in the fall of 2001 at age eighty-one.

Roger Vaughan's talent and penchant for adventure stories endures. He has written extensively and most passionately about the sea and sailing, which appear to be his personal equivalent of Piantanida's extreme skies. He lives and writes at his home on Chesapeake Bay. Vaughan's book *Fastnet*, the story of a tragic 1979 ocean race beset by a violent storm that took the lives of fifteen sailors, ends with words that recall Nick, simple words that may offer a profound explanation: "He was in his element. He was alive."

Janice Piantanida—now Janice Post—lives in the woods of the Hudson River

Valley in upstate New York with her third husband, David. Immediately following the devastating events of 1966, Janice struggled to regain her own identity and to reinvent her life. She found solace most consistently in her religious faith. Some of the things she'd witnessed at Hennepin General Hospital and at the National Institute of Health eventually led her to pursue two associate degrees and a career as an occupational therapist. "After what I'd been through, my compassion would come out," she says. "I think I understood it better than most. When I worked with cancer patients, I wouldn't make myself up and go in all fancy-dancy. Because when people are down and out, they don't really want people that are all sunshiney and have everything going great. Those people can't relate. But I could."

Vern Piantanida is retired from a successful thirty-three-year career at IBM. In his "retirement" he teaches computer science full-time at SUNY Ulster College, plays full-court basketball several times a week, and referees high-school football and basketball games (which he's done for forty-three years, always donating any money earned to charity). He has all of Nick's energy and drive without the overwhelming ego. He has entered and finished twenty-two consecutive New York City marathons—running and table tennis were the only two athletic endeavors at which Vern could always beat Nick—and recently traveled to Angel Falls in Venezuela to see for himself the face of Devil Mountain where his brother once found glory. "If I don't live to be at least 100," Vern says with a smile, "I'm really going to be ticked off."

Final Analysis

In the fall of 1970, a little more than four years after Nick Piantanida's final flight, three bright, idealistic young people lifted off from Long Island in a gas balloon called the *Free Life*. They intended to make the first aeronautical crossing of the Atlantic Ocean and by so doing to inspire and delight the world. But the truth is, they never had so much as a prayer. Only one of them had any real experience piloting balloons, they were ill-equipped and ill-prepared for the journey, and their gondola and balloon designs were horribly flawed. While it's clear that they should never have left the ground, the three were chasing their own magnificent dream, one they'd lived with for several years. They were obsessed and they were committed, despite having no real appreciation for what it would take to survive a transatlantic flight. They were never heard from again.

Piantanida, in contrast, had done a great deal to help assure the success of his project. He was neither a daredevil nor a fool. He prepared well and he did something not all adventurers are willing to do: he studied. According to Jacques Istel: "Nick read every appropriate scientific and engineering manual he could lay his hands on." He did much right on Strato-Jump. It should be remembered that the respected mid-century military balloon programs (Manhigh, Excelsior, Strato-Lab) all had their close calls, those moments that could have gone either way. Kittinger was within minutes, perhaps seconds, of asphyxiating during the descent of *Manhigh I*. Clifton McClure nearly died of heat exhaustion on *Manhigh III*. Victor Prather perished shortly after landing in *Strato-Lab V*. Each of these incidents is attributable to human error, and each might have been avoided with more procedural discipline or greater attention to detail. Not even the most well-funded, well-planned projects have bottomless bank accounts or unlimited time. Compromises, as well as mistakes, are inevitable. And, as with most things, luck comes into play. While the disaster aboard Strato-Jump III was almost certainly of Nick's own making, the problems on the earlier two flights that kept him from achieving his objectives and that necessitated the fatal third flight were clearly not. The unexpected failures of equipment from reliable, experienced contractors—the Litton balloon and the Firewel disconnect—were completely beyond Nick's control. If the Litton balloon had held together, or if the disconnect had functioned as expected, Nick might never have had to launch on May Day.

Starting from ground zero, without resources or connections, Piantanida managed to secure the assistance of one of the country's most prominent parachutists and two of its most accomplished balloonists. If Jacques Istel or Ed Yost or Don Piccard had been convinced to take leadership roles in the project, had been in position to impose their wills, wisdom, and expertise on Strato-Jump, things might have turned out differently. It seems unlikely that, given the requisite background and project authority, any of those three men would have allowed Nick to launch in Strato-Jump III without completing the prebreathing regimen. Clearly, Nick's attitude and approach to the denitrogenization process was from the beginning altogether too casual. It is also quite possible that with expert oversight, the oxygen-line disconnect problem on Strato-Jump II could have been avoided. And perhaps they could have convinced Nick of the perils of even *thinking* about touching the helmet's pressure-visor bearing above the tropopause.

"If I'd been in charge of that operation," Istel ruminated years after the fact, "it's entirely possible that there *would* have been a wrench on that second flight." But Istel disdains the Monday-morning quarterbacks who have criticized Piantanida. "Nothing is cheaper, meaner, and more demeaning than second-guessing. I hate second-guessing as a matter of principle." Istel also admits that if asked he would have refused a request to run Strato-Jump. Not only did he lack the necessary knowledge of high-altitude operations, he had felt from the start that it was simply too dangerous. In spite of several close calls at the controls of small airplanes and a life of trusting his life to parachute hardware, Jacques Istel never stepped into a balloon of any type.

If there were strategic flaws in the development of the Strato-Jump organization, they included the lack of a qualified aerospace medical person, someone with firsthand experience in super-high-altitude manned flight, in a position of authority. John Kittelson is on the money when he asks: Who focused on the pilot?

Strato-Jump desperately needed someone with the power to ground both the aircraft and the pilot. The military manned balloon programs all had this. Nick Piantanida was the only human being ever to make it above 100,000 feet who had no boss, no superior officer, no one with even temporary veto power over the dream. Surely Nick's supreme confidence in his own invincibility, a not-uncommon trait in pilots of experimental aircraft, demanded the checks and balances that only an oversight authority could have provided. Nick's mental and physical toughness had propelled him to the top of Angel Falls in spite of disintegrating shoes and a broken thumb, but courage and perseverance simply weren't enough in the stratosphere. Strato-Jump's fundamental flaw was the lack of a command structure, a structure that could have been easily provided by either the corporate or military cultures Nick had found so stifling.

In the end, the flaw was Nick's.

As Lucy Stefan points out, highly specialized medical expertise had been readily available in Minnesota. Dr. Randy Lovelace and others at the Mayo Clinic were familiar with DCS and high-altitude operations and might have been of great assistance. In 1943 Colonel Lovelace of the U.S. Army Air

Corps made a legendary no-delay-opening parachute jump from 40,200 feet in order to investigate the effects of high-altitude exposure. But while Nick had asked for the help of a number of individuals who had the direct experience that could have helped him protect himself (Joe Kittinger, David Simons, Otto Winzen), it is doubtful that he would have been willing to concede ultimate authority to any of them.

Regardless of what might or might not have occurred had Nick been willing to relinquish control, it could never have worked that way. Nick was a force of nature, and the solo act was supreme to the end. Strato-Jump very much depended on Nick's overarching, fiercely independent quality, the quality evidenced again and again throughout the course of a too-short life. It was what got him to the top of Devil Mountain after his guide quit and went home.

The final irony is that the very quality that Nick needed to put himself in position to succeed was the same quality that ultimately sabotaged the attempt. Most everybody around Nick saw both sides of that coin from the beginning. There was nothing hidden or mysterious about it. Even if he had never conceived of Project Strato-Jump, most of Nick's associates believe it would have been something else: the land speed record, the deepest ocean dive, something "impossible."

At one point during the planning sessions for Strato-Jump III, an executive of General Electric had supposedly offered Nick a lucrative desk job. But Nick had always thought of an office as a cage. "I'd rather die," he told Janice once shortly before they were married. Before discovering that the doctors had been wrong and that he was in fact capable of having children of his own, Nick had confided his own retirement fantasy: He wanted to build a log cabin deep in the mountains somewhere and adopt fifteen kids. He never stopped looking for that someplace you couldn't follow. Nick was . . . Nick.

Vern Piantanida has never completely come to terms with this aspect of his brother's personality. For Vern, who reveres the memory of his brother as he revered him in life, Nick's pursuit of the Strato-Jump dream will always be an irresponsible act. "Nick had to prove that he could do the impossible. In that sense, he was an exhibitionist," Vern, the realist, is forced to conclude.

The shock waves from Project Strato-Jump have washed down through the decades. Nick's daughters never got the opportunity to know their father.

Donna was just barely old enough to pick up and carry Nick's helmet at the Lakewood drop zone before her parents left for Sioux Falls in April of 1966. The three young women have what their mother has shared with them, which until recent years was surprisingly little. They have old photographs, some memorabilia. They have the sanitized version of Roger Vaughan's story for *Life*. They also have a toughness and a focused pride that would seem to be—at least in part—a bequest from Nick.

That pride may also come from an appreciation of their father's dream and the belief that their involuntary sacrifice was occasioned by something noble. Nick Piantanida represents a quality we profess to admire: the unquenchable drive to exceed, to surpass, to fight on to the end. Yet the individuals who would take up that quest often make us uncomfortable. Why aren't they content to recognize the limits that constrain the rest of us? What are they trying to prove?

When pioneers return from the frontier, we revere their risk-taking, their perseverance, their courage. When they become lost in the wilderness or are defeated by the elements, we shake our heads and curse them for their foolhardiness, their irresponsibility—and yet the difference between success and failure in such ventures is often a mere hairline. In his account of the ill-fated balloonists who perished in the attempted transatlantic flight of the *Free Life*, author Anthony Smith quotes a number of official and semi-official reactions to the tragedy, both critical and laudatory. "From the beginning," reads one, "the project was merely a publicity stunt." But another argues, "Splendor and folly are so near to each other. Indeed, we can barely tell one from the other, so much alike they are, and so close the line between glorious fulfillment and crushing frustration."

Nick Piantanida's last shot at the stratosphere will be remembered, finally, as a magnificent but failed attempt to extend the front lines of humankind's advance into that ultimate frontier. It was Nick's American Dream to lead that charge. One generation removed from the old city walls of Korcula, he had claimed the New World only to find that for him it wasn't enough.

It was truly, to use Jim Winker's cautiously enlightened phrase, "an attempt of the human spirit to do something remarkable."

NOTES

Prologue: Somewhere You Couldn't Follow

xii "Not a good player": Ed Madsen, interview, Oct. 23, 2001.
"You never saw him": Fred Cranwell, interview, Oct. 23, 2001.
"You know, I don't think . . . ": Bob Taglieri, interview, Oct. 23, 2001.
"I've heard him called a rebel": Jim Lagomarsino, interview, Oct. 23, 2001.

1. "Something in the Air"

The material on the early days of American skydiving and on the Lakewood Center was provided primarily by interviews with Jacques Istel, Phil Chiocchio, Lee Guilfoyle, by "'Chute Center Lures Thrill Seekers" in the *Hudson Dispatch*, and by a number of books, among them *The Endless Fall: True Stories from an Early Sky-Diver* by Mike Swain and *Bird Man* by Leo Valentin. The "boiled dog" anecdote comes from Istel's "Blackmail and Boiled Dog." The sections on the Piantanida family and on Nick's early years in Union City rely heavily on interviews with Vern Piantanida, Janice Post, Fred Cranwell, Jim Lagomarsino, Ed Madsen, Bob Taglieri, and Bob Santomenna. Joan Doherty Lovero's *Hudson County: The Left Bank* is cited in the text and was helpful on the history of Union City and environs. Nick's own journals and other unpublished writings ("Eldorado Gold" and "The Well That Ran Dry") documenting his trips to Venezuela and his exploits in La Gran Sabana were primary sources for the Angel Falls section, but additional information was provided by Vern, by an audiotape of one of Nick's slide-show presentations, and by Marshall Spiegel's article in *Practical English*. Nick shot 8mm film of both of his Venezuelan trips, and this footage was helpful. Ruth Robertson's *National Geographic* piece was also a good general reference on the area. *The Official NBA Basketball Encyclopedia*, edited by Alex Sachare, was the principal source of information on Nat Hickey, but Vern provided additional details and corrections, as well as details on the Mount Carmel tournaments and other basketball matters. Bob Taglieri and Vern were especially valuable on Nick's college months in Dodge City. Nick's files and Janice's memories were crucial to my understanding of the period just before and after the move from Union City to Brick Town.

2 "getting around to": Janice Post, interview, Oct. 21, 2001.
9 "Perhaps the children enjoyed": Lovero, *Hudson County*, 71.
10 "There was something": Taglieri interview.
11 "Hey Nick, are you crazy?": Lagomarsino interview.
13 "Nick always played to the crowd": Ibid.
"They were flamboyant": Cranwell interview.
14 "When Eddie Madsen's open": Madsen interview.
"And we fought a lot": Ibid.
"I never had the discipline": Ibid.
"Nick was the greatest guy": Ibid.
"Some business connected with speed": Ibid.
"He could be a difficult guy": Lagomarsino interview.
15 "Nick ad-libbed his life": Cranwell interview.
"I almost made sergeant": Chew, "Jumping 22 Miles from a Balloon," 1.
18 "this weirdly beautiful high jungle": Robertson, "Jungle Journey to the World's Highest Waterfall," 655.
19 "On the way to Ilse's house": Piantanida, "Angel Falls Expedition Log: 1956."
20 "It began raining last night": Ibid.
"It rained again last night": Piantanida, "Angel Falls Expedition Log: 1956."
21 "I decided nothing would stop me": Piantanida, "The Well That Ran Dry."
22 "Mating calls of a countless number": Piantanida, "Eldorado Gold."
"I strained my ear to the ground": Piantanida, "Second Expedition."
23 "Your Guides": Piantanida and Trufino, "Angel Falls Expeditions."
"You can't tell what moves you": "Travel 8,000 Miles to See Giant Falls That Were Dry?"
24 "This monster, this god": Taglieri interview.
25 "But what if I need to take": Ibid.
26 "Many people are itching": Audiotape, ca. winter 1959/60. [JP]
"If Nick said, 'Come on'": Taglieri interview.
27 "Nick and my father": Ibid.
28 "But what did Nick do": Vern Piantanida, interview, March 9, 1999 (hereafter Piantanida interview).
30 "He was the kind of guy": Bob Santomenna, interview, Oct. 23, 2001.
"Nick, we'll just sew": Ibid.

30 “I rarely dated”: Post interview, Oct. 20, 2001.
31 “Of course, Nick knew”: Ibid.
“He was the most handsome man”: Post interview, Oct. 21, 2001.
“From the day that we met”: Ibid.
32 “Don’t. You don’t want to see this”: Post interview, March 9, 1999.
33 “Call an ambulance”: Ibid.
“I trust in Nicky’s judgment”: Evans, “Project Strato Jump,” 13.
“She was in the loop”: Roger Vaughan, interview, Aug. 21, 1999.
“If it’s a boy”: Nick Piantanida, letter, December 1962. [JP]
35 “When you were with Nick”: Post interview, Oct. 20, 2001.
“I need to learn to do that.”: Ibid.
“He would just take me over”: Ibid.
36 “To produce, with extensive use”: Nick Piantanida, private papers, ca. 1962. [JP]
“1. Venezuela”: Ibid.
38 “Sometimes he would take long runs”: Post interview, Oct. 21, 2001.

2. SPACE, Inc.

Nick’s logs and files, which are full of correspondence and training records related to his efforts to plan, equip, and staff Project Strato-Jump, were a particularly valuable source on the period from his move to Brick Town up to the launch of the Litton flight. Newspaper accounts augmented Nick’s correspondence on the subject of his arrest for skydiving in New York state. Tom Crouch’s *The Eagle Aloft: Two Centuries of the Balloon in America* was the principal source on Durant’s early balloon flight. Lillian D. Kozloski’s *U.S. Space Gear: Outfitting the Astronaut* provided much of the information on Mark Ridge and his efforts to make a high-altitude flight with a pressure suit. An operational manual for the David Clark Company’s A/P22S-2 suit was a valuable reference; Jack Bassick supplied additional detail and clarifications. The material on Barry Mahon’s career came from Lee Guilfoyle and Phil Chiocchio who worked with and for him, and from several film-related databases available on the internet. Litton’s bid and proposal for Strato-Jump I provided specifications for the balloon and gondola, as well as dollar figures for the company’s services. Interviews with Janice, Vern, Jacques Istel, Joe Kittinger, Don Piccard, Karl and Lucy Stefan, Jack Bassick, Tom Smith, and Joe Ruseckas were all crucial.

39 “Good bye to you”: Crouch, *The Eagle Aloft*, 147.
40 “I reasoned that”: Nick Piantanida, training log, ca. spring 1964. [VP]
41 “I’m going to break your record”: Joe Kittinger, interview, May 19, 2001.
42 “I’m not too worried”: Ibid.
47 “this isn’t just another gas-bag dream”: Nick Piantanida, letter, July 22, 1964. [VP]
“I understood that impulse”: David Simons, interview, Sept. 11, 1989.
48 “Bad acting, bad sets”: MST3K Movie Guide Review, http://www.tilt.largo.fl.us/critic/mst3k.
49 “The average instructor”: Jacques Istel, interview, March 14, 2001.
“Without the backup”: J. H. Landefeld, letter, Sept. 25, 1964. [VP]

50 "I am not attempting to discourage you": J. R. Fleming, letter, Nov. 10, 1964. [VP]
"I might mention that the Navy": Harrison Williams, letter, Aug. 4, 1964. [VP]
51 "A four foot sphere": Don Piccard, letter, Sept. 22, 1964. [VP]
52 "All the operational phases": Nick Piantanida, letter, ca. summer 1964. [VP]
53 "Please make checks payable": Ibid.
"Why did Lindbergh": Chew, "Jumping 22 Miles from a Balloon," 1.
"safer than rolling out of bed": Terlizzi, "Readies for Record Sky Dive," 1.
"Please, you have to tell him": Post interview, Oct. 21, 2001.
54 "Fun jump": Nick Piantanida, parachute logbook, Sept. 6, 1964. [JP]
"hard pull": Phil Chiocchio, e-mail, Feb. 14, 2002. [CR]
"I saw him spinning": Ibid.
"He went in near the high-tension power line": Ibid.
"Sullivan killed this morning": Nick Piantanida, parachute logbook, Sept. 7, 1964. [JP]
56 "Nick had what it took": Vaughan interview.
"Basically": Istel interview.
"Terrific!": Ibid.
"If you're going to do it": Ibid.
"It's all promotion": Chew, "Jumping 22 Miles from a Balloon," 1.
57 "100% normal": Nick Piantanida, letter, Sept. 12, 1964. [VP]
"hands off": Nick Piantanida, jump log, ca. fall 1964. [VP]
58 "more of a glider than a parachute": Pioneer Parachute Company, brochure, ca. fall 1964.
"To show that the military government": Nick Piantanida, letter, Oct. 17, 1964. [VP]
62 "sparked": Harrison Williams, letter, Feb. 26, 1965. [VP]
"famous Piccard family": Ibid.
"preclude their acquisition": Joe Ruseckas, letter, Nov. 13, 1962. [DCC]
64 "He seemed like an impressive": Jack Bassick, interview, Oct. 22, 2001.
"Finally!!": Nick Piantanida, jump log, Feb. 15, 1965. [VP]
66 "Can you believe I've got this?": Post interview, Oct. 20, 2001.
"I would never have said it": Ibid.
67 "Your résumé is quite impressive": Marion Pruitt, letter, March 15, 1965. [VP]
"satisfactory": FAA Individual Physiological Training Record, May 14, 1965. [VP]
71 "complete understanding": Nick Piantanida, letter, May 17, 1965. [VP]
"incapacitation": Frederick Fahringer, letter, June 8, 1965. [VP]
72 "As I see it": Harrison Williams, letter, May 27, 1965. [VP]
"I am sorry": Harrison Williams, letter, June 10, 1965. [VP]
"tired of their foolishness": Staff Keegin, letter, June 15, 1965. [VP]
74 "Perfectly proportioned": Don Piccard, interview, May 23, 2001.
75 "I was never a team-type person": Ibid.
76 "I was very suspicious": Karl Stefan, interview, April 14, 2002 (hereafter Karl Stefan interview).

80 "so much useless bulk": Chew, "Jumping 22 Miles from a Balloon," 1.
81 "The slightest arm movement" Nick Piantanida, training log, Aug. 12, 1965. [VP]
"If the spin of your body": Gershen, "A Daredevil's Leap from Space," 22.
"swimming": Nick Piantanida, jump log, Aug. 16, 1965. [VP]

3. Strato-Jump I: *Frustrations End*

Strato-Jump project documents and Nick's correspondence files tell much of the story of the period leading up to the first flight, and they were consulted often, as were Nick's post-flight reports and notes. Air Force and FAA reports tell much of the story of Nick's decompression-chamber runs; the original voice tapes made during Nick's second visit to the FAA in Oklahoma City were also helpful. Janice's memories, as well as those of Tracy Barnes, provided the details for the Great Balloon Race incident. Karl Stefan, Don Piccard, and Jim Winker were very helpful on the analysis of some of the problems connected with the Litton flight, and Phil Chiocchio's eyewitness account was also valuable. Jacques Istel provided details and perspective on business matters relating to SPACE, Inc. and on some of the fundraising efforts. Mike Quigley supplied the account of Nick's meeting with NASA officials in Washington. Tapes of the air/ground communication on Strato-Jump I were invaluable. A number of newspaper articles were helpful, most of them from Minneapolis–St. Paul and New Jersey papers. Phil Chiocchio provided the Santa Claus jump anecdote, and his account was confirmed by newspaper coverage.

83 "to continue the project": Nick Piantanida, letter, July 2, 1965. [VP]
84 "I would like to have you": Ibid.
"I'm sorry the Air Force": Marvin McCall, letter, July 13, 1965. [VP]
85 "not in consonance with present requirements": Charles Berry, letter, July 16, 1965. [VP]
87 "sell a development contract": Joe Ruseckas, memorandum, Sept. 1, 1965. [DCC]
"contribute new knowledge": Stanley Mohler, letter, Sept. 9, 1965. [VP]
88 "any rocket operated by the company": Exhibit to Agreement (contract), Oct. 15, 1965. [VP]
"Nick always said": Post interview, Oct. 21, 2001.
"In a mid-air collision": Chew, "Jumping 22 Miles from a Balloon," 1.
89 "too repetitious" (and subsequent dialogue at FAA): Audiotape, Sept. 24, 1965. [JP]
90 "We feel that this project": Stanley Mohler, letter, Sept. 24, 1965. [VP]
91 "I was fairly impressed": Tracy Barnes, interview, April 22, 2002.
"It would be great" (and subsequent dialogue): Post interview, Oct. 20, 2001.
92 "We called it": Ibid.
93 "He was a delightful": Lucy Stefan, interview, April 14, 2002.
94 "an odd fellow": Post interview, Oct. 21, 2001.
"a really, really lovely woman": Ibid.
95 "I think she helped Nick": Ibid.
"I promised": Ibid.

96 "I should reach the speed of sound": "Man to Plummet 20 Miles," *Minneapolis Star*, 1.
"government support": Ibid.
"We discussed other accidents": "Brick Chutist to Try Today for 118,000-Foot Free Fall," 2.
"similarly unworried": Ibid.

98 "One of the things": Karl Stefan interview.

102 "Sure is quiet up here" (and all subsequent dialogue from flight): Audiotape, Oct. 22, 1965. [JP]

104 "This was a dream life": Phil Chiocchio, e-mail, Aug. 2, 2001. [CR]
"Watch what happens" (and subsequent dialogue): Ibid., Aug. 6, 2001.

108 "It was a little too late": David Sheridan, "Where to Land? Garbage or River," 12.

109 "Not very glamorous": Post interview, Oct. 20, 2001.
"PROJECT STRATO-JUMP I": Nick Piantanida, parachute logbook, Oct. 22, 1965. [JP]
"It hit a bit of a shear": Piccard interview.
"I didn't see anything": Ibid.

110 "It could have been a faulty seam": Karl Stefan interview.
"It was very well organized": Piccard interview.
"Piantanida threw out his hands": Rainbolt, "Chutist Goes Up in High Hopes," 1.
"I couldn't begin": Sheridan, "Where to Land?," 12.
"He was so disappointed": Post interview, Oct. 20, 2001.

111 "much too early": "Balloon Deflates Chutist's Dream," 1.
"not altogether unexpected": Sheridan, "Where to Land?," 12.
"All balloons": Ibid.
"I'm bitter": "Balloon Deflates Chutist's Dream," 1.
"Future plans are for another launch": Rainbolt, "Chutist Goes Up in High Hopes," 1.
"We find it very pleasant": Sheridan, "Where to Land?," 12.
"A number of things": Rainbolt, "Chutist Goes Up in High Hopes," 1.

112 "The flight was prematurely terminated": R. L. Schwoebel, "Strato-Lab Development: Final Technical Report," iii.

115 "Is Santa all right?" (and subsequent dialogue): Phil Chiocchio, e-mail, Aug. 1, 2001. [CR]

4. Strato-Jump II: *Second Chance*

Nick's files and Janice's memories supplied much of the background for the material on the lead-up to the second flight. Raven Industries' bid and proposal supplied specifications and details. The material on Sioux Falls, and on Raven's role, relies heavily on Christine Kalakuka and Brent Stockwell's *Paul E. (Ed) Yost, Father of the Modern Hot-Air Balloon: A Few Highlights of His Life* and on Crouch's *The Eagle Aloft*, as well as on interviews with Frank Heidelbauer, Dick Kelly, Russ Pohl, Darrell Rupp, Al Tomnitz, Roger Vaughan, Jim Winker, and Ed Yost. Emily Frisby's autobiography supplied the background on her career. Interviews with Phil Chiocchio and Jacques Istel were also helpful here. The account of William Rankin's emergency bailout is from *The Man Who Rode the Thunder*. Sources for the information on historical ballooning topics and on prebreathing are too numerous to mention here, but the bibliography contains the most important titles. The information on Andreev and Dolgov comes primarily from the Russian periodical *The Aeronaut*, which was provided to me by Jim Winker and translated for me by Olga Levadnaya. The tapes of air/ground communication for the second flight were obviously crucial, as were those containing Nick's extemporaneous oral debriefing session made shortly after the flight's conclusion. General Electric's preflight planning documents and postflight report were consulted, as were numerous newspaper articles from regional papers.

118 "Raven Industries at that time": Frisby, *Dangerous Living*, 21. [CR]
"Will Rogers character": Phil Chiocchio, e-mail, Aug. 29, 2001. [CR]

120 "It appears that if the first jump": Jacques Istel, letter, Nov. 23, 1965. [VP]

122 "If six people were saying": Vaughan interview.
"Oh, I liked Ed": Post interview, Oct. 21, 2001.

123 "I considered him to be qualified": Jim Winker, interview, Nov. 20, 1999.
"This balloon shall not be operated": "Special Operations Limitations for High Altitude Balloon" (FAA), ca. January 1966. [VP]

124 "Present methods of survival": Nick Piantanida, "Project Strato-Jump II, General Information," 1, ca. December 1965. [VP]
"We believe that special free fall training": Ibid., 2.
"the sciences of the descent of objects": "Certificate of Incorporation, SPACE, Inc.," August 1965. [VP]

125 "I've spent two years getting ready": Jim Klobuchar, *Minneapolis Star*, Oct. 20, 1965, 1D.
"I know this was a real disappointment": Stanley Mohler, letter, Dec. 1, 1965. [VP]
"No one has ever heard of Nick": Chew, "Jumping 22 Miles from a Balloon," 1.

126 "I was in awe": Dick Kelly, interview, Nov. 21, 1999.
"It wasn't a place": Ibid.
"It was a pretty rough joint": Kent Morstad, interview, May 21, 2001.

131 "balloonatic years": Frisby, *Dangerous Living*, 22. [CR]
"This is not a bizarre stunt": Chew, "Jumping 22 Miles from a Balloon," 1.
"just like floating": "Shore Parachutist Presents Plans for a 24-Mile Jump," 1.
"I guess I'll find out": Ibid.

133 "Here is a guy destined": Kelly interview.

136 "As we ascend": Karamysheva, "Record Jump," 19.
"The whole body aches": Ibid.

137 "I guess I have to go overboard": Klobuchar, 1D.

138 "Nothing's going to happen": Istel interview.

141 "savage pain": Rankin, *The Man Who Rode the Thunder*, 141.

142 "the chokes": John Paul Stapp, interview, Oct. 18, 1999.

143 "one dead son of a bitch": Joe Kittinger, interview, Sept. 17, 1991.

147 "My Dearest Darling": Nick Piantanida, letter, Feb. 2, 1966. [JP]
148 "Dearest Donna": Ibid.
151 "Tell Keuser he owes me" (and all subsequent dialogue from flight): Audiotape, Feb. 2, 1966. [JP]
163 "I could see it": Post interview, Oct. 21, 2001.
"I think he said": Ibid.
165 "We thought the gondola": Russ Pohl, interview, Nov. 20, 1999.
167 "It kind of shook me up a bit": Nick Piantanida, audiotape, Feb. 3, 1966. [JP]
"I'm sure Nick *felt*": Winker interview.
"No two of those things": Ed Yost, interview, June 26, 1999.
168 "I had to hold myself": Nick Piantanida, audiotape, Feb. 3, 1966. [JP]
174 "really pissed": Post interview, Oct. 21, 2001.
"Where's Janice?": Ibid.
"Well, honey": Ibid.
"He was never down for long": Ibid.
175 "I didn't know what to expect": Vaughan interview.
176 "I'll get up there": Andersen, "Piantanida to Make New Free-Fall Try," 1.
"I considered cutting the hose": Ibid.
177 "What was it like": Post interview, Oct. 21, 2001.
"I couldn't care less": Ibid.
"It *was* an exciting moment": Ibid.
178 "probable": Andersen, "Piantanida to Make New Free-Fall Try," 1.
"Notify VIP's of new record": Nick Piantanida, pilot checklist, January 1966. [VP]

5. Strato-Jump III: *JADODIDE*

The account of the final flight relies on the voice tapes as well as on all the sources listed for the previous chapter, as well as on interviews with Bill Jolly, Darrell Rupp, John Kittelson, Ben Raiche, Earl Clifford, and Kent Morstad. Jolly's report on the second flight, along with General Electric's report, were also useful here. Jacques Istel provided the details on his pancreatic disorder and his medical treatment. Raven's bid and proposal for the third flight were helpful. Nick's own parachute logbook provided the detail on his final jumps at Lakewood. Jim Lagomarsino provided the anecdote about Nick's visit to Union City prior to Strato-Jump III. The report on Tom Gatch's flight was provided to me by Jim Winker. Barry Mahon's film *The Angry Sky* was a useful resource throughout, but particularly for the material on the final flight. Various regional newspaper articles were also consulted.

179 "Parachutist Nicholas J. Piantanida's": Lamberto, "Balloon Flight Over Iowa Highest Ever; Jump Fails," 1.
"And there it was shimmering": F. C. Christopherson, "Of a Daring Breed to Whom We Owe Much."
181 "extremely stable": Sendler, "First Plastic-Film Parachute Jump," 20.
"The performance of the balloon carrier": Ed Yost and Nick Piantanida, letter, Feb. 4, 1966. [VP]
182 "Clearly it is better": Ibid.

182 "Abrasion marks were found": Nick Piantanida, "Preliminary Report: Strato-Jump II," February 1966. [VP]
"unheard of": Andersen, "Piantanida to Make New Free-Fall Try," 1.
"*Be certain* that all fittings": Hanff and Pegues, "Aviators Oxygen," 20.
183 "Rate of climb from launch": Stiles, "Summary Report: Project Strato Jump," 4.
"head-first to airstream": J. Faust, "Updated Project Stratojump Trajectories," G.E. press release, March 15, 1966. [VP]
187 "By that time we had three children": Post interview, Oct. 21, 2001.
"He was always trying to talk": Frank Heidelbauer, interview, Nov. 20, 1999.
188 "cheated death": "Piantanida Dies; Buddy Sets S.F. Jump in December."
"All-American breakfast": "Piantanida Is Making Plans for New Record Jump Try," 8.
"We made mistakes": UPI wire-service report, Aug. 29, 1966.
"He was gung-ho all the way": Yost interview.
189 "Dear Ed": Nick Piantanida, letter, April, 1966. [VP]
190 "Gee, Nick": Lagomarsino interview.
194 "observation tube": Nick Piantanida, audiotape, Feb. 3, 1966. [JP]
195 "This includes Istel": Ibid.
196 "Why shouldn't I go": Spiegel, "Nick Piantanida," 8.
"barbarism": "Soviets Call U.S. Cruel in Vietnam," 1.
"were using the most cruel": Ibid.
"Bring the murderer": Ibid.
198 "How hard could it be?": Phil Chiocchio, e-mail, Aug. 29, 2001. [CR]
"It was a weird evening": Ibid.
199 "What the hell's going on?": Yost interview.
"No problem, Ed": Ibid.
200 "You ready to release" (and all subsequent dialogue from the flight): Audiotape, May 1, 1966.
"He doesn't need a woman": Andersen, "Nick Struck Suddenly by Crippling Disaster," 1.
"Oh, God, he has to make it": Vaughan, "Heroic Sky-Dive Venture," 38.
206 "whoosh": Yost interview.
"hiss": Ben Raiche, interview, June 27, 2002.
"What was that?": Post interview, Oct. 21, 2001.
"Maybe he sneezed": Post interview, March 9, 1999.
207 "I want you to go home": Post interview, Oct. 21, 2001.
"I think the stress of it": Ibid.
208 "His communication system": Darrel Rupp, interview, Nov. 20, 1999.
"Hold this!": Al Tomnitz, interview, Nov. 20, 1999.
"That sure shook me up": Heidelbauer interview.
210 "Janice": Post interview, Oct. 21, 2001.
211 "You just couldn't get": Yost interview.
"Mr. Kelly and Bill Jolly": Post interview, Oct. 21, 2001.
"Can you describe for our readers": Ibid.

6. Vigil

Janice and Vern were the primary sources for the postaccident period. Their memories are augmented by Janice's correspondence files and by medical reports from the National Institute of Health and the Veterans' Hospital in Philadelphia, and by numerous newspaper articles (only some of which are mentioned in the text). Roger Vaughan and his *Life* story were consulted. The material on the tragedy aboard *Soyuz 11* relies heavily on David J. Shayler's *Disasters and Accidents in Manned Space Flight.*

213 "I'm sorry to hear": Piantanida interview, March 9, 1999.
"Nick Struck Suddenly": Andersen, "Nick Struck Suddenly by Crippling Disaster," 1.
"'EMERGENCY' AT 57,000 FEET!": Lamberto, "'Emergency' at 57,000 Feet!" 1.

215 "We're giving him every chance": "Unconscious Piantanida Treated in Minneapolis," 1.
"It was something": Post interview, Oct. 21, 2001.

216 "One time I was holding": Ibid.
"deep pain": Edward Tarlov, letter, Aug. 23, 1966.
"slight generalized flexion": Ibid.

217 "We feel that the neurological": "Doctor Says Piantanida Could Recover Fully," 1.
"Though his project terminated": "Nick's Bold Flight," 4.
"Nick, you can do it": Post interview, Oct. 21, 2001.
"You're there!": Ibid.
"one-in-a-million": Ibid.

218 "What do you mean?": Ibid.
"It's just his eye muscles": Ibid.
"I thought I would die": Ibid.
"You'd have expected me": Ibid.

219 "She's here all the time": "Piantanida Is Still In 'Waking Coma.'"
"One minute I'm optimistic": Deckelnick, "Piantanida's Wife Shares Ordeal of His Recovery," 1.
"I probably have enough influence": Ibid.
"Ever since he was a little kid": Deckelnick, "Shore Chutist Shows Signs of Recovery," 1.
"We have not actually sat down": Ibid.

220 "That poor guy": Ibid.
"stood a little taller": Tom Piantanida, letter, May 1966. [JP]

221 "Coach Moriarty has always told us": Unidentified Kansas City high-school student, letter, May 1966. [JP]
"possibly a little lighter": "Nick's Coma a Little Lighter," 1.
"There is no hope of recovery": Piantanida interview, March 9, 1999.

222 "grave . . . very improbable": "NIH Awaits Sky-Diver Injured in Plunge."
"It's a great catastrophe": "Skydiver May Live Long Yet He May Never Awaken."
"observation and supportive therapy": Edward Tarlov, letter, Aug. 23, 1966. [VP]

224 "They wouldn't let me see": Post interview, Oct. 21, 2001.

225 "He has shown some improvement": Ibid.
"massive encephalomalacia": "Hospital Summary/Autopsy Protocol," Aug. 30, 1966. [VP]

226 "You okay?": Post interview, Oct. 21, 2001.
"Just seeing those flowers": Ibid.
"Oh, Nicky looks so beautiful!": Ibid.

227 "He died of a broken heart": Ibid.

228 "If you can drive here": Vern Piantanida, letter, July 22, 2002. [CR]
"no big deal": Piantanida interview, March 9, 1999.
"But when a doctor": Vaughan, "Heroic Sky-Dive Venture," 37.
"I thought he was a good guy": Vaughan interview.

229 "his suit suddenly": Vaughan, "Heroic Sky-Dive Venture," 34.
"I was pretty upset": Vaughan interview.
"Your cares float away!": *Life*, May 13, 1966, 115.
"I sort of dropped out": Post interview, Oct. 21, 2001.
"It was too much": Ibid.

230 "It was just hard to believe": Ibid.

7. "Magnificent Failure"

Interviews with officials at the David Clark Company, Jack Bassick and Tom Smith in particular, were extremely helpful with the postaccident investigations. Dick Sears's written report on David Clark's role on Strato-Jump III was crucial, as was John Flagg's press release summarizing the company's investigation into the pressure suit and helmet, and the operational manual for the suit. Raven's press release exonerating David Clark's equipment was also important. This chapter relies heavily on a number of additional interviews, including those with Janice, Vern, Roger Vaughan, Ed Yost, Bill Jolly, Russ Pohl, Al Tomnitz, Phil Chiocchio, Darrell Rupp, Jacques Istel, Jim Winker, Darrell Rupp, Don Piccard, Earl Clifford, and John Kittelson. Joe Kittinger's book, *The Long Lonely Leap*, was consulted (and is quoted in the text), as was Anthony Smith's *The Free Life* (also quoted). The brief section on personality, genetics, and environment relies on Matt Ridley's book *Genome: The Autobiography of a Species in 23 Chapters*. Several other documents are cited directly in the text, among them the *Parachutist* article on the planned follow-up to Strato-Jump and the *Scientific American* article on contemporary high-altitude balloon/parachute projects. The film *The Angry Sky* and the voice tapes of the final flight were obviously important in the analysis of the accident. Dan Quinn's theory about Nick's tendency to strike the side of his helmet with his hand was related by Debbie Foster.

231 "gasp": "Piantanida Wages Fight for Life after Mishap," 1.
"as though he had been hit in the stomach": Richard Sears, "Strato-Jump III—Trip Report," 2. [DCC]
"anguished grunts": Horan, *Parachuting Folklore*, 152.

231 "screams": Offman, "Terminal Velocity," 133.
"He was still on his way up": Ibid.
232 "Why are we doing it?": Chew, "Jumping 22 Miles from a Balloon," 1.
233 "went off exceptionally smooth": Sears, "Strato-Jump III—Trip Report," 1. [DCC]
"Marv" (and subsequent dialogue): Ibid., 2.
234 "That's a damned lie!" (and subsequent dialogue): Ibid., 3.
"Of course, we were very sensitive": Tom Smith, interview, Oct. 22, 2001.
"the visor on his helmet": "Parachutist Is Injured Critically in 57,000-Ft. Gondola Descent," 1.
"He wore a pressurized flight suit": "Piantanida Wages Fight for Life after Mishap," 1.
"the suit showed no breaks": Ibid.
235 "equipment trouble": "Chutist Injured in Gondola Fall after Oxygen Cutoff," 40.
"Apparently": Sellick, *The Wild, Wonderful World of Parachutes and Parachuting*, 59.
"a seal on his pressure suit": Payne, *Lighter Than Air*, 271.
"the patient suffered": Edward Tarlov, letter, Aug. 23, 1966. [VP]
"visor": Sears, "Strato-Jump III—Trip Report," 2. [DCC]
"pieces of the visor": Ibid.
"That's impossible": Ibid., 3.
236 "a grave injustice": Ibid.
237 "Engineers and technical personnel": "No Indication of Piantanida Equipment Failure," Raven Industries press release, May 2, 1966. [DCC]
"Nick's face was exposed": Sears, "Strato-Jump III—Trip Report," 5. [DCC]
"All known information": Ibid., 4.
238 "1. The visor": "Notice," David Clark Company press release, May 5, 1966. [DCC]
"We will continue to investigate": Ibid.
"With all I have to worry about": Post interview, Oct. 21, 2001.
239 "Nick was too careful": Ibid.
"As a contractor": Yost interview.
"Inspections of Piantanida's space-style suit": UPI wire-service report, Aug. 29, 1966.
240 "Goddammit, he flipped his visor": Vaughan interview.
"oral/nasal breathing cavity": Bassick interview.
"left-hand bearing": Ibid.
241 "I came away from it": Vaughan interview.
242 "I remember the speculation": Rupp interview.
"Others speculated": Horan, *Parachuting Folklore*, 153.
"puddle": Audiotape, Feb. 2, 1966. [JP]
"He [Nick] commented": Tomnitz interview.
243 "According to Dan": Debbie Foster, interview, Oct. 23, 2001.
"using a trick": Ottley, "Nick Who?" 29.
"Nobody on this earth": Post interview, Oct. 20, 2001.
"On May 1, 1966": "Project New Excelsior," 2. [CR]

244 "Ed Yost said": Sears, "Strato-Jump III—Trip Report," 4. [DCC]
"overpressurized": Piccard interview.
"They say": Ibid.
"tingling sensation": Sears, "Strato-Jump III—Trip Report," 4. [DCC]
"That's classic paresthesia": Bassick interview.
"itching, usually around the ears": J. R. Brown and Melchor J. Antunano, "Medical Facts for Pilots," June, 1966, 3.
"divers' itches": Ibid.

245 "Dumping air out": Bassick interview.
"Nick had the pressure suit on": Pohl interview.
"Breathing 100-per-cent oxygen": Gantz, ed., *Man in Space*, 184.

246 "Nick starts pre-breathing": Martin Lueders, "Operations Flight Plan: Project 59561," Oct. 12, 1965.
"All during this period": Kittinger with Caidin, *The Long, Lonely Leap*, 160-61.

247 "You lose it": Bassick interview.
"Usually pilots prebreathe": Smith interview.
"You had to be there": Ibid.
"He did pre-breathe": Sears, "Strato-Jump III—Trip Report," 5. [DCC]

248 "for 30 minutes prior": Brown and Antunano, "Medical Facts for Pilots," 2.
"They weren't there": Yost interview.
"I assisted Marv": Sears, "Strato-Jump III—Trip Report," 2. [DCC]

249 "Even though he didn't have": Ibid.
"He'd reportedly done it": Bassick interview.

250 "He was *taught* to do that": Bill Jolly, interview, Oct. 23, 2001.
"Negative": Audiotape, Feb. 2, 1966. [JP]
"I couldn't open the face mask": Audiotape, Feb. 3, 1966. [JP]

251 "If he was properly trained": Bassick interview.
"*CAUTION*": "Operation and Service Manual; Fitting, Donning, Operating and Maintenance Instructions; High-Altitude Full Pressure Flying Outfit A/P22S-2," 44. [DCC]

253 "Nick, Dick Keuser said" (and subsequent dialogue): Audiotape, May 1, 1966.

256 "He would be away from me": Post interview, Oct. 21, 2001.

257 "There are requirements": Smith interview.
"Nick was always": Pohl interview.
"The flaw was this": Vaughan interview.

258 "He was always": Ibid.

259 "He had a great sense": Yost interview.

260 "If only I'd had": Audiotape, Feb. 2, 1966. [JP]
"Undoubtedly, their efforts": Kenneally, "Taking the Plunge," 24.

261 "sustenance . . . vision . . . entrusting": StratoQuest, 2001, http://www.stratoquest.com.
"deviant places of refuge": Le Grand Saut, 2002, http://www.legrandsaut.org.
"seek out, test and improve": "Project Strato-Jump—General Information," Sept. 8, 1965. [VP]

261 "1. To establish": Ibid.
262 "It has to be looked upon": Winker interview.
"PIANTANIDA 4": Drumheller, "Something about Nick," 4.
263 "Examining the history": "Project New Excelsior," 2. [CR]
"It certainly was a sobering experience": Bassick interview.
264 "you simply don't have enough": Winker interview.
265 "I think he thought it out": Yost interview.
"The only problem": Ibid.
"When you help somebody": Ibid.
"He was the kind of guy": Ibid.
"He seemed confident": John Kittelson, interview, May 21, 2001.
266 "Good launches": Ibid.
267 "Don is very interested": Karl Stefan interview.
269 "It was an eerie moment": Phil Chiocchio, e-mail, Aug. 29, 2001. [CR]
"He was in his element": Vaughan, *Fastnet*, 184.
270 "After what I'd been through": Post interview, Oct. 20, 2001.
"If I don't live to be": Piantanida interview, Oct. 23, 2001.
271 "Nick read every appropriate": Chew, "Jumping 22 Miles from a Balloon," 1.
272 "If I'd been in charge": Istel interview.
"Nothing is cheaper": Ibid.
273 "I'd rather die": Post interview, Oct. 20, 2001.
"Nick had to prove": Piantanida interview, March 9, 1999.
274 "From the beginning": Smith, *The Free Life*, 286.
"Splendor and folly": Ibid., 287.
"an attempt of the human spirit": Winker interview.

BIBLIOGRAPHY

Several of the unpublished print sources listed here and in the notes belong to private collections, the owners of which are identified as follows: Vern Piantanida (VP), Janice Post (JP), David Clark Company (DCC), and Craig Ryan (CR). In addition to print sources, a number of audio- and videotape sources are cited in the notes. The owners of the original tapes (with the exception of the audiotape of the final Strato-Jump flight, the original of which is presumed to be destroyed) are listed as appropriate in the notes using the same abbreviations given for unpublished print sources above.

Andersen, Harl. "Nick Struck Suddenly by Crippling Disaster." *Sioux Falls Argus-Leader*, May 2, 1966, 1.

———. "Piantanida to Make New Free-Fall Try." *Sioux Falls Argus-Leader*, Feb. 3, 1966, 1.

"Balloon Deflates Chutist's Dream." *Asbury Park Evening News*, Oct. 23, 1965, 1.

Bates, James M. "Early Competition Parachuting." *Parachutes Yesterday/Today/Tomorrow* (May/June 1992).

"Brick Chutist to Try Today for 118,000-Foot Free Fall." *Asbury Park Evening News*, Oct. 19, 1965, 2.

Brown, J. R., and Melchor J. Antunano. "Medical Facts for Pilots." AM-400-95/2, FAA Civil Aeromedical Institute, Aeromedical Education Division, June 1996.

Caidin, Martin. *The Silken Angels*. New York: Lippincott, 1964.

Chew, Peter T. "Jumping 22 Miles from a Balloon." *National Observer* (Jan. 24, 1966): 1.

Christopherson, F. C. "Of a Daring Breed to Whom We Owe Much." *Sioux Falls Argus-Leader*, Feb. 6, 1966, sec. A.

"'Chute Center Lures Thrill Seekers." *Hudson Dispatch*, March 1, 1965.

"Chutist Injured in Gondola Fall after Oxygen Cutoff." *New York Times*, May 2, 1966, 40.

Crouch, Tom D. *The Eagle Aloft: Two Centuries of the Balloon in America*. Washington, D.C.: Smithsonian Institution Press, 1983.

Deckelnick, Gary. "Piantanida's Wife Shares Ordeal of His Recovery." *Asbury Park Evening Press*, May 10, 1966, 1.

———. "Shore Chutist Shows Signs of Recovery." *Asbury Park Evening Press*, May 3, 1966, 1.

DeVorkin, David H. *Race to the Stratosphere: Manned Scientific Ballooning in America*. New York: Springer-Verlag, 1989.

"Doctor Says Piantanida Could Recover Fully." *Sioux Falls Argus-Leader*, May 3, 1966, 1.

Drumheller, Ed. "Something about Nick" (letter to editor). *Parachutist* (June 1966): 4.

Evans, Hal. "Project Strato Jump." *Parachutist* (April 1966): 11–16.

Frisby, E. M. *Dangerous Living: A Professional Autobiography*. Self-published, 1989. [CR]

Gantz, Lt. Col. Kenneth F., ed. *Man in Space: The United States Air Force Program for Developing the Spacecraft Crew*. New York: Duell, Sloan and Pearce, ca. 1959.

Gershen, Martin. "A Daredevil's Leap from Space." *Newark Sunday Star-Ledger*, April 25, 1965, 22.

Hanff, G. E., and R. J. Pegues. "Aviators Oxygen." *Lockheed Field Service Digest* (February 1963): 4-47.

Harford, James. *Korolev*. New York: John Wiley and Sons, 1997.

Horan, Michael. *Index to Parachuting, 1900-1975: An Annotated Bibliography*. New York: Garland, 1977.

———. *Parachuting Folklore: The Evolution of Freefall*. Richmond, Ind.: Parachuting Resources, 1980.

Istel, Jacques Andre. "Blackmail and Boiled Dog." *Athol Daily News*, Feb. 13, 1984.

Kalakuka, Christine, and Brent Stockwell. *Paul E. (Ed) Yost, Father of the Modern Hot-Air Balloon: A Few Highlights of His Life*. Oakland, Calif.: Balloon Publishing Company, March 1999.

Karamysheva, N. "Record Jump." Translation by Olga Levadnaya. *The Aeronaut: Journal of the Russian Federation of Aeronautics*, no. 2(12) (1998): 18–20.

Kenneally, Christine. "Taking the Plunge." *Scientific American* (November 2001): 24.

Kirschner, Edwin J. *Aerospace Balloons: From Montgolfiere to Space*. Blue Ridge Summit, Pa.: Aero Publishers, 1985.

Kittinger, Capt. Joseph W., Jr., with Martin Caidin. *The Long, Lonely Leap*. New York: E. P. Dutton, 1961.

Klobuchar, Jim. Untitled column. *Minneapolis Star*, Oct. 20, 1965, 1D.

Kozloski, Lillian D. *U.S. Space Gear: Outfitting the Astronaut*. Washington, D.C.: Smithsonian Institution Press, 1994.

Lamberto, Nick. "Balloon Flight Over Iowa Highest Ever; Jump Fails." *Des Moines Register*, Feb. 3, 1966, 1.

———. "'Emergency' at 57,000 Feet!" *Des Moines Register*, May 2, 1966, 1.

Lovero, Joan Doherty. *Hudson County: The Left Bank*. Sun Valley, Calif.: American Historical Press, 1999.

"Man to Plummet 20 Miles." *Minneapolis Star*, Oct. 16, 1965, 1.

National Aeronautic Association of the USA. *World and United States Aviation and Space Records*. Arlington, Va.: NAA, 2001.

"Nick's Bold Flight." *Sioux Falls Argus-Leader*, May 3, 1966, 4.

"Nick's Coma a Little Lighter." *Sioux Falls Argus-Leader*, May 8, 1966, 1.

"NIH Awaits Sky-Diver Injured in Plunge." *Asbury Park Evening Press*, June 16, 1966, 10.

Offman, Craig. "Terminal Velocity." *Wired* (August 2001): 128–35.

"Operation and Service Manual; Fitting, Donning, Operating, and Maintenance Instructions; High-Altitude Full Pressure Flying Outfit A/P22S-2." David Clark Company (September 1962). [DCC]

Ottley, Bill. "Nick Who?" *Parachutist* (October 1978): 24, 29.

Overs, Donald E. "Light Heart Trans-Atlantic Balloon Flight Attempt: An Analysis." AIAA 7th Aerodynamic Decelerator and Balloon Technology Conference, San Diego, Oct. 21-23, 1981.

Payne, Lee. *Lighter than Air: An Illustrated History of the Airship*. New York: Orion Books, 1977.

Piantanida, Nick J. "Angel Falls Expedition Log: 1956." Unpublished journal. [VP]

———. "Eldorado Gold." Unpublished article, ca. 1960. [VP]

———. "Second Expedition." Unpublished journal, ca. 1959. [VP]

———. "The Well That Ran Dry." Unpublished article, ca. 1960. [VP]

Piantanida, Nick, and Rudy Trufino. "Angel Falls Expeditions." Brochure, ca. 1959. [VP]

"Piantanida Dies; Buddy Sets S.F. Jump in December." *Sioux Falls Argus-Leader*, Aug. 30, 1966, 1.

"Piantanida Is Injured Critically in 57,000-Ft. Gondola Descent." *Minneapolis Tribune*, May 2, 1966, 1.

"Piantanida Is Making Plans For New Record Jump Try." *Asbury Park Evening Press*, April 1, 1966, 8.

"Piantanida Is Still in 'Waking Coma.'" *Asbury Park Evening Press*, May 8, 1966, 8.

"Piantanida to Make New Free-Fall Try." *Sioux Falls Argus-Leader*, Feb. 3, 1966, 1.

"Piantanida Wages Fight for Life after Mishap." *Asbury Park Evening Press*, May 2, 1966, 1.

"Project New Excelsior." Unpublished planning document, ca. early 1990s. [CR]

Rainbolt, Richard. "Chutist Goes Up in High Hopes." *Minneapolis Star*, Oct. 22, 1965, 1.

Rankin, William H. *The Man Who Rode the Thunder*. New York: Pyramid Books, 1961.

Ridley, Matt. *Genome: The Autobiography of a Species in 23 Chapters*. London: Fourth Estate Limited, 1999.

Robertson, Ruth. "Jungle Journey to the World's Highest Waterfall." *National Geographic* (November 1949): 655–90.

Sachare, Alex, ed. *The Official NBA Basketball Encyclopedia*. 2nd ed. New York: Villard Books, 1994.

Schwoebel, R. L. "Strato-Lab Development: Final Technical Report." General Mills, Inc., Dec. 31, 1956.

Sears, Richard. "Strato-Jump III—Trip Report." David Clark Company, Worcester, Mass., May 1966. [DCC]

Sellick, Bud. *Skydiving*. Englewood Cliffs, N.J.: Prentice-Hall, 1961.

———. *The Wild, Wonderful World of Parachutes and Parachuting*. Englewood Cliffs, N.J.: Prentice-Hall, 1971.

Sendler, P. A. "First Plastic-Film Parachute Jump." *Sky Diver* (May 1966): 20.

Shayler, David J. *Disasters and Accidents in Manned Spaceflight*. Chichester, England: Praxis Publishing, 2000.

Sheridan, David. "Where to Land? Garbage or River." *Minneapolis Tribune*, Oct. 23, 1965, 10.

"Shore Parachutist Presents Plans for a 24-Mile Jump." *Asbury Park Evening News*, Jan. 20, 1966, 1.

"Skydiver May Live Long Yet He May Never Awaken." *Pensacola Journal*, June 2, 1966, 1.

Smith, Anthony. *The Free Life*. Wainscott, N.Y.: Pushcart Press, 1995.

Smith, J. R. "Proposal 3031: Balloon Program in Support of Project Strato-Jump." Raven Industries, Inc., Sioux Falls, S.D., Nov. 15, 1965.

———. "Proposal 3190: Balloon Program in Support of Project Strato-Jump III." Raven Industries, Inc., March 9, 1966.

"Soviets Call U.S. Cruel in Vietnam." *New York Times*, May 2, 1966, 1.

Spiegel, Marshall. "Nick Piantanida." *Practical English* (April 28, 1967): 7–8.

Stefan, Karl. "Technical Proposal 2523: Project Strato-Jump." Applied Science Division, Litton Systems, Inc., St. Paul, Minn., June 29, 1965.

Stiles, R. L. "Program Plan: Project Strato-Jump." General Electric, Philadelphia, Jan. 17, 1966.

———. "Summary Report: Project Strato-Jump." General Electric, Philadelphia, Feb. 24, 1966.

Swain, Mike. *The Endless Fall: True Stories from an Early Sky-Diver*. Pittsburgh: CeShore Publishing, 2000.

Terlizzi, James Jr. "Readies for Record Sky Dive." *Hudson Dispatch*, Sept. 10, 1965, 1.

Tobin, Francis E. Untitled space and aviation column. *Worcester Evening Gazette*, ca. early June 1966.

"Travel 8,000 Miles to See Giant Falls That Were Dry?" *Hudson Dispatch*, ca. January 1959.

"Unconscious Piantanida Treated in Minneapolis." *Sioux Falls Argus-Leader*, May 2, 1966, 1.

Vaughan, Roger. *Fastnet: One Man's Voyage*. New York: Seaview Books, 1980.

———. "Heroic Sky-Dive Venture." *Life* 60 (May 13, 1966): 32–39, 115.

Valentin, Leo. *Bird Man*. London: Hutchinson, 1955.

Von Braun, Wernher, Frederick I. Ordway III, and Dave Dooling. *Space Travel: A History*. New York: Harper and Row, 1985.

INDEX

Bold indicates pages with tables or illustrations.